the sustainable home

Cathy Strongman

the sustainable home

the essential guide to eco building, renovation and decoration

MERRELL
LONDON · NEW YORK

Contents

Introduction

Anyone attempting to build a sustainable house in the late 1990s would have faced an uphill struggle. Neighbours would have fretted over the arrival of a 'wacky' building on their street, suitable technology and materials would have been hard to find and expensive to buy, and the majority of architects would have balked at the thought of compromising the aesthetic integrity of their designs to cut carbon emissions. Today the scenario could not be more different. Environmental concerns have risen from the bottom to the top of the world's agenda, national governments now campaign on green issues, and concerns about sustainability have infiltrated almost every aspect of life.

The home – the place where people spend most of their time and invest the majority of their money – has become central to the drive to halt climate change and secure the future of the planet. Conscientious clients are now demanding more sustainable homes, and product designers, engineers, planners and architects have responded to this challenge. Sustainable houses that are visually exciting and financially attainable have sprung up in countries across the world. Where sustainable architecture was previously a fringe movement, now it is at the cutting edge of design.

Sustainability is a complex word that is hard to define and even more challenging to quantify. It was first brought into popular use in 1987 in the United Nations document *Our Common Future*, a report by the World Commission on Environment and Development, which defined sustainable development as meeting 'the needs of the present without compromising the ability of future generations to meet their own'. Since then the term 'sustainability' and such expressions as 'green', 'eco' and 'environmentally friendly' have been used to describe a multitude of products and actions that show concern for the earth's

resources. The overuse of these terms has threatened to reduce the potency of their meaning.

At the most basic level, a sustainable house is one that has a significantly lower impact on the environment than a standard building. Two key strategies prevail: reducing the amount of energy needed to construct the building in the first place, and minimizing the building's reliance on energy once it is occupied. The former might be achieved by selecting materials that require little energy to extract and produce. Referred to as having a 'low embodied energy', these are often natural substances, such as timber and clay. The sustainable home might also employ renewable, salvaged or recycled components that do not deplete the world's natural resources, or it might incorporate local materials, thereby reducing the carbon emitted during transportation and creating contextual buildings that express the local vernacular. Materials must also be employed economically: if smaller quantities are used, then less energy is needed for their extraction, production and transportation. Chemical finishes should be avoided, as these require large amounts of energy to produce and release harmful fumes, degrading air quality. The impact on the site should be kept to a minimum, thereby helping to support existing biodiversity and reducing the need for intrusive foundation work. This might be achieved by preventing the unnecessary demolition of existing buildings, or by reducing the waste created during construction.

In terms of cutting energy consumption in the finished building, it is important to include thermally massive materials, which store and release heat energy gradually over time, so that the house can naturally regulate its internal temperatures. Similarly, insulation helps to keep a house cool in the summer and trap heat inside in the winter. Solar gain can be exploited through extensive south-facing glazing, while protective shades offer a passive means to prevent the house from becoming uncomfortably hot during the warmer months. Such an approach will also make the best use of natural daylight. Low-energy light fixtures should be used when artificial light is required, and all appliances should be energy-efficient. Natural ventilation can be provided through manually operated

windows and doors or through an integrated ventilation system, while water-collection systems can harvest rainwater or treat grey water from sinks and showers for reuse in gardens or toilets. Forms of renewable energy, such as solar panels and wind turbines, can also be incorporated. The design should ensure that the building will have a long life by creating a flexible and comfortable home where people can continue to enjoy spending time, but it should also allow for the safe dismantling and recycling of the building after use.

A sustainable house will incorporate a variety of these methods, but the exact combination depends on the context of the project and the desires of both the architect and the client. Whether the aim is to build a house from scratch, renovate an existing building or simply spruce up interiors, there are now hundreds of options available to create a more sustainable home. The first section of this book is aimed at offering solutions for a broad range of circumstances, budgets and ambitions, and includes everything from paint to insulation. Products and installation methods are explained, as well as the environmental advantages and disadvantages of each.

The thirty houses in the second section of the book provide inspirational examples of how architects and clients have put these methods to the test. Found in locations across the world, from Russia to Hawaii, they demonstrate that building ecologically sound homes no longer means compromising on style. The projects vary in scale, from a garden studio in Canada to the Swiss Ambassador's home in Washington, D.C. Some are placed on tight urban sites, others in suburbs and yet others in remote rural settings. While several projects readdress materials and techniques that have existed for centuries, others explore cutting-edge technologies and exploit new materials in unique ways.

These houses provide a thrilling snapshot of the rapidly developing field of sustainable architecture. Equally importantly, they demonstrate the enormous range of approaches that are now being developed to create houses that coexist in harmony with our planet. There is no single optimum design for a sustainable home: the possibilities are endless and the prospects unbelievably exciting.

Materials and Techniques

The essential guide to sustainable building and decorating materials and techniques, from light bulbs to house-building, and from cushions to new flooring. Each is fully described, with special instructions for its use or application and a list of advantages ⊕ and disadvantages ⊖, helping you make an informed choice to suit your circumstances.

Construction Materials

Timber

Living trees extract carbon dioxide from the atmosphere and convert it into carbohydrate to sustain themselves, giving out oxygen as a by-product. Even after felling, timber continues to act as a carbon sink, as the carbon absorbed by the tree during its lifetime is only released once the timber is burnt. In addition, it is renewable, recyclable and easy to reuse, and expends minimal energy in its production. When used effectively in construction it will last hundreds of years, longer than the time a new tree will take to grow.

The most environmentally friendly timber available is second-hand, recycled or reclaimed wood from reclamation centres or architectural salvage yards. The next best alternative is local timber that comes from a sustainably managed forest or plantation. Buying locally cuts down transport pollution and helps sustain the local economy, as well as producing a site-sensitive home, since local materials are likely to have been used in the area for generations. For sustainable wood, highly regarded certification schemes are operated by the Forest Stewardship Council (FSC), the Programme for the Endorsement of Forest Certification (PEFC), the Canadian Standards Association (CSA) and the Sustainable Forestry Initiative (SFI).

Method: Traditional timber frames

The majority of timber-framed houses are made from a softwood skeleton, hidden by plasterboard on the inside and cladding on the exterior. In some cases, as in oak-framed buildings, the beams are left exposed. A typical timber-framed house can be put together on site by a small team within a matter of weeks. Such buildings lend themselves to high levels of insulation because only the frame bears the structural load, leaving the rest of the wall cavity free for insulation.

Method: Timber-panel systems

There are now many prefabricated timber-panel systems available. They reduce construction time on site and, because of their enhanced strength, offer greater flexibility of design. Steko blocks, for example, are large, hollow timber blocks that slot together to form load-bearing walls. No glue or other fixings are necessary and a whole house can be built in three days. Some panel systems, such as structurally insulated panels (SIPs), also incorporate insulation.

Method: Timber cladding

Numerous species of timber can be used to clad a building. Larch, Western red cedar, Douglas fir and sweet chestnut are all good examples of stable and hard-wearing cladding materials that will not shrink or warp. When left untreated these timbers will naturally turn a silvery grey. Also available are such products as ThermoWood, treated with heat to drive out moisture and resin, making them durable and stable. Air gaps should be left between timber boards to allow walls to breathe and prevent the wood from rotting.

+
- Renewable and recyclable.
- Will retain carbon dioxide from the atmosphere during the building's life.
- Timber-framed buildings are easy to adapt, making them viable for life-long occupancy.
- Timber construction is economical and fast, and less messy than wet construction methods such as bricks and mortar.

−
- Illegal logging is still commonplace, so it is important to know the source of the timber.
- Timber buildings will last a long time only if they are constructed properly: a good vapour barrier and protection from splashback at ground level are essential.
- Low thermal mass: unlike heavy masonry materials, timber cannot store and release radiant heat gradually. This can cause rapid fluctuations in temperature in poorly designed lightweight timber-framed houses, leading to a reliance on mechanical forms of temperature control.

The timber frame is one of the oldest building methods.

Timber-panel systems are quick and simple to use.

...rmoWood cladding uses heat-treated timber for strength and durability.

...can be sculptural and decorative as well as functional.

Cob

Cob is made from sand, straw and clay-based subsoil mixed with water to produce a moist and malleable building material. It has been used for centuries – the word cob originates from the Old English for lump or rounded mass – and is non-toxic, non-exhaustible and 100 per cent recyclable, making it extremely ecological.

Method

Cob was traditionally mixed by stamping the ingredients together, but it can also be combined by hand or mechanically. Lumps or 'gobs' of the mix are then pressed together to form walls. Building with cob is slow, since each layer of 450–600 millimetres (18–24 inches) takes three or four days to dry.

- Outstanding thermal qualities.
- Highly durable, resulting in buildings that will last for hundreds of years.
- Made entirely from natural and recyclable materials, consuming virtually no energy and producing no pollution in manufacture.
- Malleable: cob can be shaped into expressive sculptural forms.
- The soil from which cob is constituted can often be dug from the site, reducing costs and eliminating transport pollution.

- Extremely thick walls, impractical for small plots or tight corners.
- Must be built on a masonry foundation to protect against rising damp, and gutters, drainpipes and an overhanging roof must be properly maintained to prevent excessive moisture penetrating the walls.
- Slow to build with.

Unfired Clay Bricks

Unfired clay bricks have been produced for centuries all over the world. They are now being mass-produced in factories, where they are dried naturally by air, reducing the embodied energy of the product.

Method

Unfired clay bricks are suitable only for non-load-bearing walls unless the weight of the building is carried by a timber or steel frame. Typically they are used for partition walls, built using clay mortar and a clay or lime plaster.

- Made entirely from natural and recyclable materials, consuming virtually no energy and producing no pollution in their manufacture, unfired clay bricks have only 14 per cent of the embodied energy found in fired bricks.
- Inexpensive.
- Good acoustic properties.
- High thermal mass helps to regulate the indoor temperature of a building.
- Hygroscopic: absorb and release water, allowing walls to breathe, which regulates interior humidity levels and prevents damp.
- Particularly appropriate in the restoration of old buildings where new materials need to match existing natural materials.

- Suitable only for non-load-bearing walls.
- Specialist skills are needed to produce a high-quality clay-plaster finish that will prevent the blockwork from deteriorating.

Unfired bricks have low embodied energy.

Construction Materials

Fired Bricks

Fired bricks were invented in 3500 BC and have become the world's most common form of masonry. They are made from a natural material – clay – but have a high embodied energy because they are fired at temperatures of 900–1200°C (1650–2190°F). The most environmentally friendly option is to buy reclaimed bricks, although it is essential to match the type of brick to its function: facing bricks for cladding, strong engineering bricks for structural walls, and common bricks, which are durable but not attractive, for foundations and internal walls.

Method

Bricks come in a bewildering array of designs, shapes, textures and colours, and they can be laid in various patterns, called bonds. They can be used for cladding, framework, foundations and features such as chimneys. Roughly 17 per cent of a brick wall is made up of mortar, which holds the bricks in place. Portland cement is the most common mortar material, but traditional lime mortar is increasingly being revived, as it is made from a natural non-toxic material and allows walls to breathe.

+
- Made from a natural material; most clay pits are carefully managed, and refilled and replanted after use.
- Highly durable and low-maintenance.
- High thermal mass.
- Can be reused or recycled as construction aggregate.

–
- As a result of the firing process, bricks have a high embodied energy.
- Most often laid with Portland cement, which is also a high-energy product.
- Often used as exterior cladding on lightweight buildings, which negates their thermal-mass properties, since any radiant heat escapes outwards into the atmosphere.

Fired bricks are one of the most familiar building materials.

Straw-Bale Construction

Americans have been constructing houses out of straw bales for hundreds of years, and this material has now taken off in other countries. Straw is a renewable and natural material that requires little energy to extract and process. It is also cheap and easy to use. Grown in abundance in many parts of the world, it can often be bought from local farmers or agricultural merchants.

Method

The original 'Nebraska' technique involved stacking the straw bales on top of one another to provide support for the roof. It is now more common to build a post-and-beam frame and stuff the walls with the bales for insulation. Lime, clay and cement can be used for the exterior cladding, or more progressive materials such as corrugated steel or clear polycarbonate, revealing the straw bales behind.

+
- Highly insulating.
- Fire resistant: a plastered straw-bale wall will act as a firewall for at least two hours.
- Good acoustic performance.
- Breathable walls: when combined with a natural plaster, bales allow a gradual transfer of air through the wall, bringing fresh air into the living environment.

–
- Poorly made bales are liable to collapse; if the bale holds together when lifted by one string, it is sound.
- Bales must be kept dry: moisture content of over 15 per cent can cause fungal and bacterial growth that will weaken the structure.
- Straw can attract insects and vermin. It is advisable to wrap the bales in a layer of anti-insect mesh.

Straw bales are easy to use, but must be kept dry.

Hemp

Hemp is fast becoming a major force in sustainable construction. It is renewable and natural and can be grown without herbicides or insecticides, producing up to 10 tonnes per hectare (4 tonnes per acre) per year and improving the condition of the ground. Hemp stems contain long, strong fibres that can be processed to produce two materials – hurds (the woody core) and fibres – both used in the construction industry. Because of the speed at which it grows, the minimal energy used in its processing and the plant's ability to act as a carbon sink, hemp has the potential to produce carbon-negative construction materials.

Method

The list of hemp's building applications is impressive and continues to grow. Such products as fibreboard, roofing tiles, wallboard, panelling and bricks can be made from compressed hemp hurds. The hurds can also be mixed with a combination of lime products to produce a lightweight and airtight insulating material that resembles cement. This can be moulded into load-bearing bricks, poured into wooden shutters and set as external and internal walls, or sprayed as lime plaster on to interior and exterior surfaces. Interior walls can be left exposed or finished with a natural paint, but exterior walls should be finished with a lime render. Hemp bales can also be inserted into a timber frame in a similar way to straw bales.

+
- Provides good thermal and acoustic insulation.
- Resistant to rotting, rodents and insects.
- Despite being a lightweight material, hemp is surprisingly durable. It is currently being explored as a reinforcement material for concrete.
- Hygroscopic: hemp can absorb and release moisture, allowing the building to breathe.
- Hemp lime is favoured among restoration and conservation specialists, as it is well suited to old stone buildings.

−
- Hemp is not currently grown in large quantities, so products are more expensive than conventional materials.
- Grows only in certain countries, so it may be difficult to find a local source.

Concrete

Consisting of cement, aggregate (most commonly gravel and sand), water and chemical admixtures, concrete is a material that divides the construction industry. While detractors point to its exceptionally high embodied energy (cement alone is responsible for 8 per cent of global carbon dioxide production), supporters highlight life cycle assessments that show that concrete buildings are durable and need less energy to run than lightweight buildings. The debate will continue, but the concrete industry is working to improve its ecological credentials. Recycled aggregates such as glass are increasingly being used as partial replacements for gravel and sand, as are manufacturing by-products such as ground, granulated blast-furnace slag (GGBS), created during iron production, and pulverized fuel ash (PFA), also known as fly ash, a by-product of coal-fired power plants. These alternatives can constitute up to 30 per cent of the mass of structural concrete and 70 per cent of that of non-structural concrete, cutting greenhouse gas emissions by up to 40 per cent.

Method

Concrete can either be pre-cast in a factory or cast *in situ*. Pre-cast concrete comes in a number of forms, including flat-panel systems, modular units and foundations, beams and floors. *In-situ* concrete is generally poured between wooden shutters, which are removed after it has set, although it can be poured between permanent insulating boards made from such materials as expanded polystyrene.

+
- Can be cast in any shape and size, offering great flexibility in design.
- High thermal mass.
- Exceptional strength and durability.
- Good acoustic properties.
- Can be recycled into construction aggregate.

−
- High embodied energy.
- Produced from non-renewable materials.
- More messy and cumbersome than dry construction methods such as timber frames.

This concrete bears the marks of its shutter moulds.

Construction Materials

Prefabricated houses minimize construction damage to the surrounding area. This is the M-House, which is supplied in two pieces and joined together on site.

Prefabrication

Prefabrication has been hailed as more energy-efficient, less polluting through transport, less material-wasting and more enduring (parts can be reused) than on-site construction. In reality, the boundary between prefabrication and on-site construction is not straightforward: all builders apply some form of prefabrication, whether it be in the form of a nail or an insulated sandwich board. Prefabrication is not always the most environmentally friendly solution: a factory that throws away its off-cuts is less green than an on-site builder who uses his on the next job, and a local construction company using local materials will generate less transport pollution than a prefabricated house that is transported hundreds of miles by ship or lorry.

Method

It is important to choose a prefabrication manufacturer that has high ecological standards in terms of materials, the technology it can incorporate within the house, and its general attitude towards such factors as waste management and transport. Ideally, the company should be local, with an employment and material pool close to the site.

+
- Mass production of parts reduces the cost of prefabricated houses.
- Can be planned, built and inhabited within months.
- Built to exact specifications, with less left to chance during the construction process.

–
- Still carries some negative connotations in some countries because of bad practice in the past (post-war British prefabs are a good example), so might be harder to sell.
- Can never be truly bespoke.

Specially coated glass reduces glare and heat gain.

Adjustable solar shades can be a striking design feature.

Glass

Glass lets in natural sunlight as well as radiant heat, which, when harnessed properly, can help to heat a house. But careful positioning is essential: glazed areas can create uncomfortably high internal temperatures in the summer. Buying poor-quality windows and glass doors is a false economy, since they lose heat, resulting in higher heating bills.

Method: Improving existing windows

About 20 per cent of the heat in a house is lost through single glazing. Some of this can be saved by upgrading existing windows with secondary glazing. The cheapest solution is to attach a clear plastic film (polythene) to the inside of the window with double-sided sticky tape. The most effective and long-lasting option is to fit a second window inside the existing one; this is still cheaper than replacing the window entirely.

Method: Installing new windows

The best option is double- or triple-glazed, low-emissivity (low-e), argon-filled windows. Low-e coatings and argon both improve the thermal performance of a window by reducing heat flow between the panes of glass. A reflective or semi-reflective coating on the outside of the window will reduce radiant heat transfer into the building and cut out the sun's glare (although reflectivity should be below 20 per cent in order not to annoy neighbours). Timber frames (preferably approved by the FSC) use far less energy in their manufacture than aluminium or PVC frames. Specialist companies can provide 'conservation glazing' for historical buildings or design high-performing replicas.

Solar shades

Well-designed solar shades will block out excessive summer sun while allowing lower-lying winter sun to penetrate windows. Attached to the exterior of the building, they can be either permanent fixtures or mechanical devices that rely on electricity but can be retracted fully in the winter.

+
- Glass brings natural light into the home.
- Can introduce radiant heat into a building when positioned correctly.
- Windows, doors and skylights can be opened to ventilate a building naturally.

–
- Poor-quality windows leak heat.
- Glass has no thermal mass and poor insulation qualities compared to solid walls.

Construction Materials

Metal

Numerous metals are used in the construction industry, steel and aluminium being the most common. Such metals are generally lightweight, strong and durable and can contain up to 95 per cent recycled material, while being themselves recyclable. They are, however, high-energy products, which always use some non-renewable materials in their production. Zinc has the lowest embodied energy of all metals.

Method

Steel frames and studwork are often used in buildings because component parts are prefabricated, lightweight and allow thin walls, maximizing internal floor area. Such metals as zinc, steel and aluminium are also used as a cladding material for walls and roofs, as well as for smaller details such as window frames.

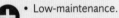

- Low-maintenance.
- Flexible: metals can be moulded into dynamic shapes.
- High strength-to-weight ratio.
- Most metal construction products are totally recyclable.

- High embodied energy.
- Metal is still often produced using virgin materials.
- Most steel is made from iron ore mined in Australia and Brazil, and could create considerable pollution through transport.

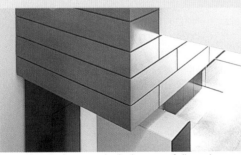

Zinc has the lowest embodied energy of all metals.

Green roofs are gaining in popularity all the time. They maintain valuable habitats as well as providing excellent insulation.

Green Roofs

Where environmental performance is concerned, green roofs considerably outstrip any other roofing material. They replace vegetation covered or destroyed by buildings and encourage biodiversity, provide excellent sound and heat insulation and absorb water run-off, making them an effective tool against flash flooding. They are increasingly being incorporated into sustainable buildings: since 2000 more than 30 million square metres (323 million square feet) of green roof have been installed in Germany alone.

Method

Most domestic green roofs consist of a waterproof underlay covered with soil or crushed stones and either topped with seed mats or planted with seeds. Sedum, a low-growing succulent plant, is popular because it is resistant to wind, frost and drought and requires little maintenance, but a wide variety of plants can be grown as long as the building is strong enough to support the load. Using local substrates as the growing medium and planting local seeds will support local biodiversity, as will including such objects as dead wood to provide extra habitat for wildlife.

- Extended roof life: a green roof protects the waterproofing membrane from climatic extremes, UV light and mechanical damage, doubling its life expectancy.
- Provides thermal insulation in the winter and helps to cool buildings in the summer.
- Excellent sound insulation.
- Growing mediums can be made from recycled materials, such as crushed bricks.
- Reduces water run-off.
- Encourages biodiversity.
- Provides attractive green spaces.

- Where the roof pitch is steeper than 40 degrees, a green roof is unfeasibly expensive.
- Plant growth is unpredictable because of airborne seeds and those dropped by passing birds.

Insulation

Installing a good layer of insulation in the walls and roof of a house is a simple and extremely effective way of improving the building's environmental performance. An insulated building will suffer less from heat loss and will therefore need less energy to run. Unfortunately, the materials with the best insulating properties have historically been damaging to the environment: plastics converted into foam using ozone-depleting gases, for example, are good insulators. As manufacturers have been forced to replace these gases (such as CFCs) with non-harmful alternatives, the performance of the resulting insulation has diminished. Some man-made insulators, such as cellular glass, which is made from recycled car-windscreen glass, are still highly effective, yet these are still high-energy products. The greenest option is to use natural materials, such as hemp, coconut fibre, cellulose fibre or sheep's wool.

Method

Insulation should be chosen in response to the structure of a house: sheep's wool, for example, is good for timber-framed houses with ventilated walls but cannot be used in wet masonry walls. Within these limitations, it is important to choose materials that are organic, from a renewable source, produced in an eco-friendly way and, wherever possible, local. Good examples are sheep's wool, cellulose (made from recycled paper), flax, hemp, corkboard, strawboard and wood fibre. Some materials, such as Misapor concrete, have insulation built into them. For houses with thin walls, products such as structurally insulated panels (SIPs) contain petrochemical-based thermal insulation, which has a high embodied energy but performs extremely well.

 • The more effective the insulation, the less energy is consumed.

• Some insulation materials have a high embodied energy.

Sheep's wool is a natural and safe insulating material.

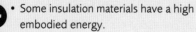
A water butt: the simplest water-management system.

Water Management

At a time when climate change is causing droughts to occur more frequently, investing in a rainwater-harvesting system can greatly reduce mains water consumption. Systems range in complexity, cost and effectiveness. A state-of-the-art model includes filters, pumps, storage tanks, an integrated rainwater-harvesting unit and environmentally friendly guttering and will supply all domestic needs except for drinking water. At the other end of the scale, a simple water butt – a storage unit attached to a drainpipe – is cheap and will provide enough water to irrigate the garden and wash the car. Grey water (from showers, baths, washing machines and dishwashers) can also be collected in a tank, filtered and reused to flush toilets, although not for bathing or drinking. Most commercial filters use a charcoal, cellulose or ceramic cartridge that must be cleaned or replaced regularly, and there are also systems that filter the water through sand or plants, which absorb pollutants into their roots.

Method

Basic systems such as water butts are easy to install, but for more complex systems a specialist is needed. The design of a building – for example, the shape of its roof – can affect the amount of water that can be harvested, so water management should ideally be taken into account by an architect at the design stage.

• Reduced consumption of mains water results in financial savings.
• Rainwater is soft, so in hard-water areas it will make appliances last longer.
• Using grey water will maintain supply for longer during droughts.

• Initial outlays can be high, and for more complex systems the payback period is likely to be ten to fifteen years.
• Water storage tanks take up space and are often ugly. They can, however, be concealed underground or within the building, or made into a feature in themselves (as in the Cape Schanck House, pages 68–73).

Renewable Energy

Ground-Source Heat Pumps

Ground-source heat pumps extract latent heat from the ground via a borehole or network of underground pipes. The heat is transferred to a mix of water and antifreeze carried in the pipes and then converted, via the heat pump, into thermal energy for heating and hot water. Technically this is not renewable energy, as electricity is needed to power the pump, but that electricity amounts to one-third of the total energy produced, so the excess could be termed renewable. The pump can also be powered by solar energy or another renewable form.

Method

Heat pumps are most suited either to newly built properties, where they can be buried underneath the structure, or to existing houses with garden space. It is advisable to employ a professional to carry out a geothermal assessment of the site and install the system. In straightforward cases, the installation can be completed in two days.

+
- System is easy to maintain and lasts at least twenty years.
- Lower running costs than oil, liquid petroleum gas, coal and electric heating systems.

−
- The temperature generated – 50°C (122°F) – is too low for a regular radiator to heat a room. Heat pumps work better with underfloor heating, which can achieve room temperatures of 25–28°C (77–82°F) from water temperatures of 45–65°C (113–149°F).
- Can be installed only where the ground is suitably soft for digging a trench or borehole.
- Higher running costs than mains gas heating.

Solar slates are extremely discreet photovoltaic panels.

Photovoltaic Panels

Photovoltaic (PV) panels consist of two or more thin layers of semi-conducting materials that, when exposed to light, generate electrical charges, which are conducted by metal contacts as direct current into a building's electricity supply. The most efficient are monocrystalline panels, which are cut from single silicon crystals and harness 15 per cent of the sun's energy. Positioned on a south-facing roof, 5 square metres (54 square feet) of top-grade monocrystalline panels can generate around 600 kilowatt-hours over a year – about 20 per cent of an average family's needs. Multicrystalline panels, made from melted silicon cut into wafers, are slightly cheaper and still harness 13 per cent of the sun's energy.

The industry is rapidly developing, and more discreet forms of PV panel are appearing on the market. Solar slates, designed to replace conventional roof tiles, are neat and effective. Sharp has developed the Lumiwall, which looks by day like a tinted window but contains solar-powered LEDs that light up at night. Copper indium

A ground-source heat pump uses heat energy from the ground to provide heating and hot water.

gallium selenide (CIGS) is also being explored as a replacement material for silicon. Silicon panels are produced in a similar way to computer chips but CIGS is printed on thin, flexible polymer sheets and can be applied directly to roofs, windows and cladding.

Method

PVs are suitable for any wall or roof that faces within 90 degrees of the sun's midday point and that is strong enough to carry the extra weight, and they require very little maintenance. A contractor should install the panels, as the electrical connections can be complicated. It is cheaper to fit a solar-power system during construction of a house, when scaffolding is already in place, than during a refit.

- Produce no waste or pollution once installed.
- Low-maintenance with a lifetime in excess of twenty years.
- Fixed cost for electricity as opposed to the increasing cost of mains power.
- Reliable and constant energy source: the sun rises every day and PVs work even in overcast or rainy weather.

- Silicon is an expensive material and supplies are limited. Solar panels are therefore unlikely to fall considerably in price.
- Although they work all year round, there is a considerable difference in energy supply between the summer and the winter.

These solar panels are mounted flush with the roof, and provide hot water for bathrooms and kitchen in a family house.

Solar Thermal Hot-Water Panels

Roof-mounted solar thermal panels are a straightforward and proven means of harvesting the sun's energy to provide free hot water. They come in two designs. Evacuated-tube systems are the more advanced, and comprise glass pipes containing small amounts of antifreeze hermetically sealed within a small central copper pipe. When heated by the sun, this antifreeze converts to steam, which rises to the top of the tube, transfers its heat to a collector head and condenses back into liquid for the process to be repeated. Flat-plate systems comprise water-carrying copper pipes bonded to a copper absorber plate. The fluid is heated by the sun and transported to an insulated water tank, where the heat energy is transferred to the domestic water through a heat exchanger. A typical system consists of 3–4 square metres (32–43 square feet) of panels mounted on a south-facing roof, feeding into a 200-litre (44-gallon) tank. This will provide around 30 per cent of an average household's hot-water needs (more in the summer months). A hot-water cylinder stores the water heated during the day so that it can be used beyond sunlight hours.

Method

The system can be mounted on any roof that will bear the load, and is compatible with all boiler systems except combination boilers. Professional installation is recommended.

- Little maintenance is required.
- An average system will reduce a household's annual carbon emissions by 350 kilograms (770 lb).
- Can be used for larger applications, such as swimming pools.

- Most systems are guaranteed for only five to ten years.
- Expensive to install.

Renewable Energy

Biomass

Domestic biomass systems vary in complexity from a stand-alone stove that will heat a single room to a wood-pellet boiler with an automated fuel supply that can run a central-heating system or heat water. For domestic use, the fuel is usually wood pellets, wood chips or logs, although non-wood biomass fuel, such as animal waste, biodegradable products from the food-processing industry and such high-energy crops as rape, sugar cane and maize, can also be used. The ecological merit of biomass fuel is disputed by some, as it still pumps carbon dioxide into the atmosphere when it is burnt. But the system is effectively carbon neutral, since the carbon dioxide released is equal to that absorbed by the tree during its lifetime, and a biomass boiler can save 6–7 tonnes of carbon dioxide a year compared to fossil-fuel-generated electricity.

Method

Qualified stove dealers can advise which appliance will suit the circumstances and help install a safe and efficient system. Installation is straightforward, and the boilers are often compatible with existing radiators, pipe networks and underfloor heating. Wood systems must be cleared of ashes, and a regular fuel supply is required, so they are most appropriate for homes close to a wood supplier and with plenty of storage space.

- Cheaper than electricity, liquid petroleum gas and oil.
- Wood-burning stoves are more efficient than open fires, and logs burn slowly so the stove requires little attention.
- Wood ash, unlike coal ash, is an excellent fertilizer.
- There are many attractive stoves available in both traditional and contemporary styles.

- More expensive than mains gas.
- Wood pellet boilers make a whirring noise.
- Requires storage space, although wood pellets are a much denser energy source than logs.
- Combustion releases carbon dioxide into the atmosphere.

A wood-burning stove is comforting as well as ecological.

Wind turbines are slowly becoming a more common sight.

Domestic Wind Turbines

Wind turbines use the wind's force to rotate aerodynamic blades that turn a rotor, creating electricity. Most small wind turbines generate direct current (DC) electricity and require a battery and an inverter to convert DC into alternating current (AC) electricity. If the system is connected to the national grid, the conversion is done automatically and no battery is required. Domestic wind turbines are relatively cheap and easy to install but are currently not a viable option for many homes: optimum results are achieved at wind speeds of 12.5 metres (41 feet) a second, whereas the average wind speed in built-up areas is 4.6 metres (15 feet) a second; and they must be installed on a high mast (at least 6.5 metres/21 feet) to avoid the unusable turbulent air that passes over the top of a building.

Method

Wind turbines can be mounted either on a free-standing mast or on the roof of a house. They should be installed only in areas where wind speeds regularly reach 6 metres (20 feet) a second or more and where there are no obstacles such as hills, trees and other buildings that are likely to reduce wind speed or increase turbulence. Their visual impact and noise also need to be considered. Ideally, a professional assessment should be carried out before a domestic wind turbine is purchased.

- Wind power is clean and renewable.
- Produces no carbon emissions or waste products once installed.

- Dependent on the speed and direction of the wind, which is highly unpredictable and varies according to location, the height of the turbine and nearby obstructions. Most homes are not in the most suitable sites.
- Usually not cost-effective.
- In very strong winds a turbine could damage the structure of the building to which it is attached.
- Highly visible.

Heat-Recovery Systems

Domestic heat-recovery systems consist of two separate air-handling units: one collects and exhausts stale air from the inside of a building; the other draws in fresh air and distributes it around the home. Both airstreams pass through a heat-transfer module and, although they remain physically separate, heat from the exhaust air is transferred to the fresh incoming air. The system also works in reverse, so that in the summer the air being exhausted from the building cools incoming warmer air. This reduces the energy needed to heat a house in the winter and cool it in the summer.

Method

Small wall-mounted heat-recovery systems are available, but the majority are centralized, with the heat-recovery unit installed in the loft and attached to ducts positioned throughout the house. A central panel controls the fan speed and the temperature of the air entering the building. For best results the system should be serviced annually, by either the owner or a professional, a task that involves cleaning or replacing the filters and vacuuming inside the ducts.

- Improves interior air quality by removing pollutants and odours.
- Uses the heat energy generated by washing machines, showers, cookers and other human activities.
- Obviates the need for noisy extractor fans or unsightly trickle ventilators in windows.

- Fans in the system require energy to run, so it is less energy-efficient than natural ventilation in the summer.

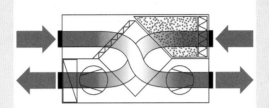

A heat-recovery system supplies warmed, fresh air.

Micro-Combined Heat and Power

Combined heat and power (CHP) units capture and use the heat generated when fuel is burnt to produce electricity. CHP technology can be applied on various scales, from a single home to a large industrial plant. Micro-CHP units are suitable for use in the home, replacing the conventional boiler in a central-heating system with a small gas engine that drives an electrical generator. This powers domestic lights and appliances, while the waste heat from the engine can heat both rooms and water.

CHP units harness wasted heat to generate electricity.

Method

Installation should be carried out by a trained and certified professional. Homes fitted with CHP are usually also connected to the mains electricity grid and may retain back-up boilers to ensure that they are never short of an energy supply during maintenance of the CHP plant or during periods of unusually high energy requirement. Maintenance is straightforward: the heating system within the CHP unit requires servicing in the same way as any residential boiler; the generator machinery requires a routine service similar to an automotive tune-up every eighteen months to two years. The majority of systems use natural gas for fuel, although CHP is compatible with most forms of fuel, including biomass sources.

- Approximately 90 per cent of fuel is converted into useful heat and power (compared to 70–80 per cent in conventional boilers and as little as 50–65 per cent in older boilers), meaning lower fuel consumption, reduced energy costs and lower carbon emissions.
- Often generates electricity that is surplus to demand and can be sold back to the utility company.
- Secure and reliable supply of both electricity and heat energy.

- A new technology that is in development; manufacturing costs and thus retail prices are currently high.

Floor Surfaces

Timber floors are natural and attractive, and are best installed over a sound-absorbent underlay.

Natural-Fibre Carpets

A wide variety of carpets are made from natural plant fibres including hemp, seagrass, jute, paper, coir and sisal. Each has a distinctive appearance and texture: seagrass has a slightly shiny finish, and sisal is the most hard-wearing. Many of these natural plant fibres can be combined with one another or with wool. They also come in a wide variety of weaves and patterns and some, such as sisal, can be naturally dyed.

Method
Carpets can be nailed or glued to an underlay made from hessian, jute or natural latex. Low-VOC, water-based and formaldehyde-free glues are available.

+
- Rich texture and natural earthy tones.
- From a renewable source.
- Low embodied energy.

−
- Often manufactured in India and East Africa so it is important to consider whether the carpet has been made in ethical conditions and how far it has travelled.
- Should last from five to seven years. More durable types are advisable in such busy areas as hallways and stairs.
- Stains easily; only dry cleaning methods are advisable, as exposure to water can cause shrinkage.

Timber Floorboards

For renovations, existing floorboards should be saved if possible. For newly laid wooden floors, the most environmentally friendly options are reclaimed floorboards or sustainably certified wood or bamboo, which is technically a grass but has a similar appearance to timber.

Method
Existing floorboards can be restored by sanding them down and restaining or varnishing them, filling any gaps with strips of wood or a sealant such as Gapseal. Floor finishes made from vegetable oils and natural waxes contain no biocides or preservatives.

+
- A well-maintained timber floor will last for decades and can be recycled or reused.
- Timber from a sustainably managed forest is one of the most ecological building materials on the market.

−
- Poor acoustic qualities unless fitted with a sound-absorbent underlay.
- Non-natural sealants contain harmful volatile organic compounds (VOCs).

Natural-fibre carpets are warm and attractive.

Concrete is increasingly being used for domestic interiors.

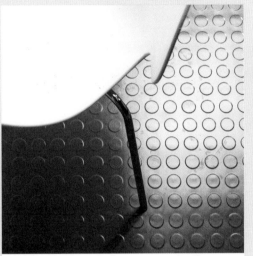
Rubber is hard-wearing but can accrue air miles.

Stone is suitable for use inside and out.

Polished Concrete

When polished, concrete becomes highly reflective. This extremely durable floor surface has gained considerably in popularity since the 1990s.

Method

For renovations, existing concrete that has been hidden under another floor covering can be polished, as can virgin concrete in a newly built house. Dry polishing produces less waste, as it gives off dust rather than sludge.

- Polishing concrete that has already been laid as part of the construction process is cost-effective and reduces the materials used.
- Easy to clean, never needs waxing or coating, and extremely durable.
- Works well with underfloor heating.
- High thermal mass: absorbs and releases heat energy gradually, helping to regulate internal temperatures naturally.

- High embodied energy.
- The polishing process is both labour- and energy-intensive.

Rubber

This renewable product is made primarily from the sap of the Pará rubber tree (*Hevea brasiliensis*). Both recycled rubber floors (made mostly from tyres) and natural rubber floors are available in many colours.

Method

Rubber floors can be laid in strips or tiles, or poured into place. There is usually no need for an adhesive.

- Renewable and recyclable.
- Water-resistant: suitable for all rooms.
- Non-toxic, sound-absorbent and extremely durable.

- Some contain PVC, plasticizers or halogens, which release toxins at every stage of their life (from production to disposal), so it is important to check the content before buying.
- Rubber is non-biodegradable.
- Most rubber is now grown in south-east Asia, so the pollution produced during transportation should be considered.

Stone

Stone floors, including slate, limestone, flagstone, granite and marble, are durable and full of character. Reclaimed or local stone is the best option in ecological terms.

Method

Stone floors can be laid over concrete or timber floors, although they are heavy so it is important in timber buildings to check whether the structure can take the weight. Stone floors are fixed in place with mortar, lime mortar being the most environmentally friendly product.

- An abundant, reusable natural resource.
- Easy to maintain.
- The most durable flooring surface available.

- Non-renewable.
- Stone quarrying can degrade the landscape.
- Extremely heavy.

Floor Surfaces

Linoleum has shaken off its 'institutional' reputation.

Tiles are endlessly varied and adaptable.

Warm and comfortable, cork is also totally renewable.

Linoleum

Linoleum flooring is manufactured from linseed oil, resin, wood flour or cork powder, and pigments. It is durable, scratchproof and easy to clean; it is also waterproof, making it ideal for bathrooms and kitchens. It is a good choice for those with asthma and allergies as it resists bacteria and is antistatic, repelling dust and pollen. It is available in a wide range of colours and patterns and can be cut to fit around irregular shapes.

Method
Can be laid as sheets or tiles. Low-VOC, water-based and formaldehyde-free adhesives are available.

- Manufactured from renewable materials.
- Recyclable and biodegradable.
- Durable, with a lifespan of more than forty years.

- Chemical fertilizers are often used in the production of linseed.
- Flammable.

Tiles

A dazzling array of colours, sizes and patterns is available. Tiles can be used in kitchens and bathrooms, to segregate an open-plan living area or for decoration in a hallway. The greenest option is to use reclaimed tiles; if new tiles must be used, those with a high recycled-glass content (up to 100 per cent) are best.

Method
Environmentally friendly adhesives are preferable to traditional cement mortar.

- Recyclable.
- Durable and easy to maintain.
- Suitable for wet areas, such as bathrooms.

- Substantial amounts of energy are consumed during firing.

Cork

Cork is a naturally produced material harvested from the bark of the cork oak (*Quercus suber*). The tree is unharmed in this process and the bark grows back over a period of nine to twelve years. Cork floors come in a range of patterns and tones, from golden yellow to a nutty brown, usually pre-sealed, and can be used in any room of the house, including high-traffic areas.

Method
Cork is laid with glue, preferably a low-VOC, water-based and formaldehyde-free adhesive.

- Renewable, recyclable and biodegradable.
- Excellent thermal and acoustic insulation.
- Contains suberin, a natural substance that resists mould, mildew and bacteria, making it a good choice for allergy sufferers.
- Affordable.
- Durable: should last at least twenty-five years.
- Low embodied energy.

- Most cork floor tiles are transported from Portugal, with concomitant pollution and energy consumption.
- Some cork floors contain formaldehyde.

Wall Coverings

Clay Plaster

A blend of clay, fine aggregate and organic fibres, clay plaster is versatile, easy to apply, cost-effective and environmentally friendly, making it a practical alternative to gypsum and cement-based plasters. A variety of colours can be achieved using natural pigments, and wall surfaces can be left exposed or covered in paint or wallpaper.

Method

Clay plaster is stored as a dry powder, to which clean water is added before use. It hardens by drying rather than through a chemical reaction, so it remains workable for a long time after application and can be reworked at any time with the addition of water. It can be applied to surfaces such as gypsum plaster and plasterboard after a layer of primer has been applied. It is not necessary to seal clay plaster surfaces except in places where they are exposed to splashing water or high humidity.

+
- Absorbs and diffuses water vapour, making it ideal for buildings in which walls need to breathe.
- Absorbs odours.
- Keeps walls cool in summer and warm in winter.
- Effective sound insulator.
- Low embodied energy.

−
- Drying time, typically a week, is longer than that of standard forms of plaster.
- Not approved by all building codes.
- Few experts for application and repair.

Lime plaster is easy to work with and flexible once dry.

Lime Plaster

Lime plaster is made from processed lime, water and aggregate. It has been used for centuries and is compatible with many traditional building materials that are becoming increasingly popular for their sustainable properties, such as stone, cob, straw bales and timber. When left exposed it has an attractive earthy appearance.

Method

Lime plaster comes in powdered form. Mixed with clean water, it is easy to apply and can be coloured using natural pigments. It is finished with a coat of limewash or distemper.

+
- Lime plaster can be reworked for days after the initial application.
- Flexible: can accommodate structural movement and is less likely to crack than modern plasters.
- Moisture-permeable, allowing walls to breathe.

−
- Made by firing limestone, which releases carbon dioxide, giving lime plaster a high embodied energy.

Clay plaster gives a natural, more earthy look than modern, artificial plasters.

Wall Coverings

Paint

Synthetic paints are made from petrochemical and mineral resources, and contain high levels of volatile organic compounds (VOCs), which continue to be released as gas months after application, contributing to atmospheric pollution. They can also cause health problems, such as allergies, asthma and skin irritation. Water-based synthetic paints are generally safer for the user and contain fewer VOCs, but this reduction is usually achieved by increasing the load of other toxic chemicals, such as ethylene glycol, which has been linked to cancer. As well as being toxic during manufacture, application and disposal, the paint contains large quantities of chemical compounds, resulting in a high embodied energy. The production process also creates a lot of waste compared to final product. Of all synthetic paints, zero-VOC, 100-per-cent acrylic emulsion is the best option.

The greener alternative is to use natural paints that contain either no VOCs at all or such naturally occurring VOCs as pine resin or citrus-based solvent at greatly reduced levels. Oil-based emulsions, such as linseed-oil paints, containing natural pigments are the most environmentally friendly choice, followed by wood- and vegetable-based resin paints, casein paint (derived from milk), mineral paints, clay paints (made from naturally occurring clay) and limewash.

Method

Most natural paints can be applied in exactly the same way as synthetic paints. Limewash, however, may need several coats.

+
- Oil-based emulsion, wood- and vegetable-based resin paints and casein paints are made from renewable resources.
- Casein paints, clay paints and limewash are non-toxic. Other natural paints are considerably less toxic than synthetic paints.
- Natural paints have a low embodied energy.
- Natural paints are microporous, allowing water to pass through without causing flaking or peeling.
- Synthetic paint ingredients are electrically charged and use plastic so they attract dust and bacteria; natural paints use plant resins and other materials that repel dust and bacteria, producing healthier environments.
- Most natural paints are totally biodegradable.

−
- Natural paints are more expensive than synthetic paints.
- Some natural paints, such as clay paints, come only in a limited range of colours.
- Natural paints take longer to dry.

Ecological paints contain natural ingredients rather than chemicals, and give a healthier home environment.

Wallpaper

The most environmentally friendly wallpapers are printed on recycled paper, sustainably produced wood pulp or natural plant fibres, such as jute or seagrass. The inks traditionally used for wallpaper emit harmful VOCs, but some companies use water-based inks or natural dyes. Alternatively, vintage wallpaper can be used.

Method

Water-soluble wallpaper pastes that are free from acrylic, solvents, fungicides, preservatives and synthetic resins are now available.

+
- Wallpaper hides imperfections in the wall better than paint, so is ideal for ecological restoration projects.

−
- Environmentally friendly varieties can be hard to find and are more expensive than standard designs.

Insulating paint is a quick way of regulating a building's heat loss and gain, where conventional insulation is not practical.

Insulating Paint

There are now on the market paints, such as Thermilate and Insuladd, that add an instant layer of insulation. They are made from insulating microspheres: tiny hollow ceramic balls developed by NASA to combat the high temperatures encountered by a space shuttle on re-entering the atmosphere. A vacuum at the centre of the microsphere reflects and refracts heat because it is non-conductive. On internal walls and ceilings, insulating paint reduces heat loss; on external walls and roofs, it reflects radiant heat, helping to keep internal temperatures cool.

Method

Insulating paint can be bought in pre-mixed form or as an additive to be mixed with any paint. It is applied in the same way as normal paint.

+
- An easy way to improve insulation instantly.
- Non-toxic.
- Can be used on a wide range of surfaces, including timber and steel pipes.

−
- Adds a slightly grainy texture to walls and, when mixed into a darker paint, lightens the colour. Both these problems can be rectified by adding a final coat without the additive.
- Cannot be mixed into plaster or wallpaper adhesive.

This paper by Louise Body is from managed forests.

Fabric

Domestic fabric is traditionally full of chemicals, such as dioxins, flame retardants and chlorine bleach. These not only emit toxins in the home but also cause pollution during their manufacture and disposal. In addition, the crops used to produce such fabric are grown using hazardous pesticides, which affect the health of workers and contaminate the soil and local water supplies.

However, the market for environmentally friendly fabric has expanded rapidly since the late 1990s. There are now many natural fibres available that have been grown with little pesticide or none at all: organic cotton, linen and silk are relatively easy to find; hemp fabric can be made into a whole range of products, including shower curtains and sofa covers; and there are also fabrics made from less obvious materials, such as cork, soya, bamboo, corn and tree bark. The German company Nettle World even sells fabric made from stinging nettles.

The increasing range of fabric made from recycled materials includes Repreve by Unifi, a 100-per-cent-recycled polyester yarn made from post-consumer and post-industrial waste, while Ting makes cushion covers from old seatbelts. Old fabric can be recycled into patchwork duvets, a new set of curtains or a lampshade – there are endless possibilities.

Method

The best way to find sustainable fabric is through website searches and speciality eco stores, since most major stores stock only a limited range. It is important to check the origin of a fabric before buying it. Adding a thermal liner to curtains will reduce heat loss through windows.

➕
- Natural fibres often come from renewable sources and consume little energy in their production.
- Fabric made from pesticide-free crops will be non-toxic, creating a healthier home environment.
- Recycling and reusing fabric saves not only on landfill, but also on the energy and resources that would have been used to create new products.
- Sustainable fabric is rarely mass-produced, and will give a home a point of difference.

➖
- Sustainable fabric is difficult to find and choice is therefore limited.
- Because it is not often mass-produced, sustainable fabric tends to be more expensive than conventional material.

Organic cottons and linens; Ting seatbelt cushions (right).

Furniture

Whether it is an antique Victorian dining table or a coat-stand saved from a skip, reusing a piece of furniture will save the energy needed to manufacture a new one and the waste generated by dumping the old. Outdated or scruffy furniture can be reupholstered or painted, recycling it into something new. Furniture makers are also now producing new objects out of recycled materials, from chandeliers dangling discarded spectacle lenses to tables put together from scraps of wood, and from chairs made of 100-per-cent recycled plastic to babies' cribs fashioned from recycled cardboard. These objects are also often themselves recyclable after use. Furniture ranges made from such natural and renewable materials as bamboo and sustainable timber are also becoming widely available, often through well-established designer brands.

Method

For new furniture, rather than buying something that is not exactly right it is often worth commissioning a bespoke piece. This allows complete control over the materials used, and the end result is likely to last longer. Second-hand furniture can be found on web-based auction sites and in antiques shops and auction showrooms. Products made from natural, renewable or recycled materials are now available from some major furniture stores, as well as websites and specialist shops. Many of the latest generation of designers are ecologically aware and are pushing the boundaries when it comes to sustainable design.

+
- Most antique furniture is well made and will increase in value with age.
- A sustainable piece of furniture, be it reused, recycled or made from responsibly sourced materials, will often be unique.
- Some of the most exciting young designers are now exploring environmentally sound materials and methods.

−
- Time and effort are required to find second-hand furniture, since most major stores do not stock pieces.
- One-off designs are more expensive than mass-produced furniture.

Recycling can be immensely stylish and original. This is the Tide *chandelier by English designer Stuart Haygarth.*

New York-based Scrapile makes new pieces from old wood.

Lighting

CFL bulbs are bright and extremely energy efficient.

Several types of energy-efficient light bulb are now available. Compact fluorescent bulbs (CFLs) are the most common, consuming about 66 per cent less energy than incandescent bulbs and lasting an average of ten to fifteen times longer. They come in a full range of whites, from warm to cool, as well as other colours, and fit most domestic sockets, as well as being compatible with dimmers. Krypton light bulbs replace the argon gas found in incandescent bulbs with krypton, which allows the bulb filament to run at a higher temperature at a lower wattage and emit a whiter light. They are also dimmable. Halogen bulbs are another energy-efficient alternative, particularly useful for smaller lights, modern track lighting and recessed wall lights. They use inert gases and a small amount of halogen, and last, on average, two to three times longer than a standard incandescent, burning at higher temperatures to give off a brighter light per watt. LED lights are good for focus lights, such as desk lamps and reading lights, as they produce minimal heat but maximum light. They do not have filaments but are powered by the movement of electrons in a semiconductor material. Although they are the most expensive bulbs, they last 133 times longer than a standard incandescent and approximately ten times longer than a CFL bulb.

Method

Incandescent bulbs should be avoided in favour of the energy-efficient alternative that is best suited to the fixture. Dimmers are an effective way of saving energy and will extend the life of halogens or dimmer-compatible CFLs. Motion detectors are also a high-tech method of ensuring that lights are in use only when there is someone in the room.

+
- Energy-efficient bulbs last longer and consume less energy, saving money and reducing carbon emissions.
- Changing a light bulb is the simplest and cheapest way to improve the environmental performance of a home.

−
- CFLs contain tiny amounts of mercury, so it is important not to inhale the vapour if they break. They must be disposed of as hazardous waste.
- Halogen bulbs get very hot, so they must be kept away from flammable items. They should be held during installation with a plastic bag over the hand, as oil from fingers will reduce the life of the bulb.
- Incandescent light bulbs are still the most widely available and the cheapest option.

Appliances

The major problem with household appliances is that they are left on standby when not in use. Every year video cassette recorders and DVD players in the UK alone use £255 million of electricity when on standby. But standby buttons are gradually being phased out and manufacturers are competing to create appliances that are more energy-efficient when in use. In the meantime, there are gadgets available that make switching off at the mains easy. Such devices as the British Power Safer and Bye Bye Standby and the German Vigor Ecosave can be inserted between an appliance and the plug socket and will automatically disconnect the device once the standby mode is activated. A new mobile-phone charger from the UK company Carphone Warehouse stops using power as soon as the handset is recharged. Alternatively, such devices as the Australian-invented Power Genie turn off all appliances in one room with a single switch. Of course, all household appliances consume electricity, and in some cases it may be worth assessing how much these items are needed.

Method

The easiest option is simply to pull the plug on televisions, stereos, phone chargers and anything else that glows when it is not in use. The alternative is a device that cancels out standby, or, for new appliances, a model without a standby function.

+
- Switching off at the mains is a simple and effective way of reducing power consumption over the lifespan of an appliance.

−
- Gadgets that prevent standby often cost more than the electricity saved, but some now control numerous appliances at once, shortening the payback period.

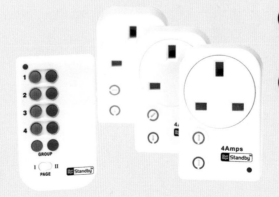

Anti-standby devices connect to multiple appliances.

Heating

Underfloor Heating

High-tech plastic pipes containing water are buried in the screed underneath a floor or run just below the surface. Because heat is transferred to the entire floor area, water temperatures of 45–65°C (113–149°F) can achieve room temperatures of 25–28°C (77–82°F); regular radiators would have to reach 80°C (176°F) to achieve a similar room temperature. If the system is installed alongside a condensing boiler, energy savings of up to 40 per cent can be made. Even with a standard boiler, 15 per cent energy savings are normal.

Method

Compatible with all floor finishes, underfloor heating can be installed throughout a house on any storey, or combined with radiators fitted in other rooms. Because it employs full lengths of piping without any joints, the system is almost maintenance free and has a lifespan of up to one hundred years.

+
- Reduces the circulation of dust and dust mites.
- Invisible and does not take up valuable wall space.
- Warmer at foot level than at head level, creating a more comfortable environment.
- Rooms can be independently controlled with individual programmable thermostats.
- Quiet.
- Compatible with alternative energy sources such as solar thermal hot-water panels (see page 23) and ground-source heat pumps.

–
- More expensive to install than radiators, although more economical in the long run.
- Long heating up and cooling down period.
- Involves considerable disruption when installed in existing buildings.

Condensing boilers capture heat from waste gases.

Condensing Boilers

Condensing boilers are the most energy-efficient boilers available. The combustion process generates gas by-products, including water vapour and carbon dioxide, and in a conventional heating system these are vented out of the house. Condensing systems cool the combustion gases, capturing the additional heat that is released and distributing it to the home. Where an older conventional boiler will run at between 60 and 70 per cent efficiency and a modern combination or conventional boiler at around 80–84 per cent, a condensing boiler will run at up to 97 per cent efficiency.

Method

Condensing boilers are widely available and as easy to install as conventional boilers. They come as floor-standing and wall-hung units and in a variety of sizes, including models small enough to fit into a kitchen cupboard.

+
- More energy-efficient, reducing energy demands and saving money.

–
- 20–30 per cent more expensive than conventional systems.

Underfloor heating is a simple system that requires much lower water temperatures to produce comfortable warmth.

Bathrooms

Low-Flow Showerheads

A typical shower uses 20–40 litres (4½–8½ gallons) of water per minute, compared to 140 litres (30½ gallons) for a full bath. But, confusingly, a power shower uses even more water than a bath. The best solution is a low-flow showerhead, which cuts the water flow to 9.5 litres (2 gallons) per minute or less. Low-flow showerheads come in two types: aerating and non-aerating. Aerating is the most popular, as it draws air bubbles into the water to give a fuller shower spray with steady pressure.

Method

Installing a low-flow showerhead is a simple way of reducing water consumption. They can be bought in most bathroom shops and are simply screwed into place. Many new showers are automatically fitted with a low-flow showerhead.

- Instant savings on water and heating bills mean a short payback period.
- Aerating showerheads are self-cleaning, as the air bubbles eliminate any build-up of mould or calcium.
- Cheap and easy with impressive results.

- Can seem disappointing compared with a power shower.
- Because water from an aerating showerhead is mixed with air, the water temperature can decrease towards the floor of the shower.
- Aerating showerheads produce excess steam.
- With non-aerating showerheads the water flow tends to pulse.

Aerating low-flow showerheads produce a satisfying spray.

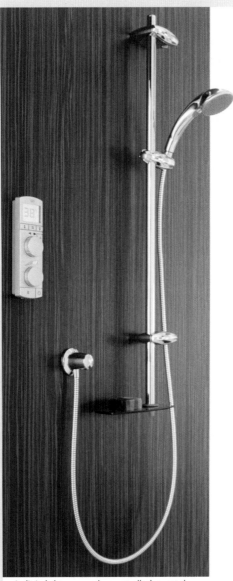

A digital shower can be controlled remotely.

Digital Showers

Digital showers are a new product that, used responsibly, can save water. They are operated by a wireless remote control, meaning that you can turn the shower on from your bed. When a pre-set temperature and flow are reached, a signal indicates that the shower is ready and reduces the flow by half until someone gets in. The shower will switch itself off after five minutes if it is not used. There is also a pause button, enabling users to save water while they lather their hair or shave.

Method

Digital showers are designed to be easy to install and to cause as little disruption as possible to the bathroom. The processing box, which mixes the water to the required temperature, can be installed away from the showering area – in a loft or airing cupboard or underneath a bath, for example. The water pipe and data cable can be concealed in the wall or riser rail.

- Discreet technology.
- Saves water normally wasted when waiting for the shower to heat up or adjusting it to the right temperature.
- Offers low flow rates and pause during showering.

- More expensive than a conventional shower.
- Requires responsible choices when setting flow and temperature.

Dual-Flush Toilets

Conventional toilets can flush as much as 13 litres (3 gallons) of water at every use, but a dual-flush toilet will use only 2–3 litres ($^1/_2$–$^2/_3$ gallon) on a short flush and 4–6 litres (1–1$^1/_2$ gallons) on a full flush. Dual-flush cisterns are now obligatory in some countries and are globally available. They come in as many styles as conventional toilets.

Method

A dual-flush toilet can be installed in the same way as a conventional toilet and should be included in all new bathrooms. An economical way to cut the flow of a conventional toilet is a Hippo – a polyethylene open-ended box that sits in the cistern. It holds back 3 litres ($^2/_3$ gallon) of water from each flush, cutting a household's water consumption by up to 15 per cent, and is incredibly cheap to buy.

- Quick payback periods of three to five years owing to impressive water savings.
- Widely available in a broad range of styles.

- More expensive than conventional toilets.
- Almost all dual-flush toilets use a flush-valve mechanism and can suffer problems from flush-valve leakage.

Baths

If they must be used at all, baths should be used sparingly. The best option is a salvaged tub or one that has a reduced water capacity, with fast-flowing taps to cut heat loss while it is filling. Bathwater can be reused to irrigate the garden.

Taps

There are many types of tap that are more efficient than conventional models. It is more economical to run a mixer tap, which combines hot and cold water, than two separate taps. Some taps have a reduced water flow. Certain ranges of eco taps have a lever control fitted with a water brake that keeps the water flow at a moderate level. Taps that work on a sensor automatically switch off when there is nothing underneath them.

Method

Mixer taps, reduced-flow taps and eco taps come in a variety of styles and are becoming increasingly widely available. They can be fitted in the same way as conventional models. The flow of an existing tap can be reduced by attaching a water-reducing valve (an aerator that will add bubbles to the tap's flow and reduce the water consumed by up to 50 per cent) or by inserting a cartridge that converts the tap's output to a spray.

- Mixer taps require only one tap rather than two, cutting down on the materials and energy used in manufacture.
- Automatic taps require less cleaning, as soap- and dirt-covered hands do not touch the tap.
- Reduced-flow taps cut water consumption, leading to savings on water and heating bills.
- Quick fixes are available that are simple to install and cost-effective.

- Automatic taps must be powered either from the mains or with batteries.
- Eco taps tend to be more expensive, although prices are becoming more competitive.

Dual-flush toilets are now obligatory in some countries.

A mixer tap uses less water than two separate taps.

Kitchens

Wooden Units and Worktops

When a kitchen is being refurbished, existing cupboard carcasses can be kept and covered with different fronts and work surfaces. MDF is energy-intensive and contains formaldehyde, but, if it must be used, a formaldehyde-free version is available. A better option is wood from a sustainable source; oak, pine and birch are common choices. Salvaged wood can also be cut down to size to make unit doors and worktops, and will often be of a higher quality and greater strength than new wood, as old wood tends to have been grown more slowly, giving it a tighter grain. An increasing variety of worktops made from sustainably sourced timber is now available.

Method

Kitchen units and work surfaces made from certified timber are becoming increasingly easy to buy, with such international furniture stores as ILVA and IKEA stocking ranges. Alternatively, a carpenter can make a bespoke kitchen with sustainably grown wood, which can be made watertight by rubbing in natural oils, even olive oil, or coated with an environmentally friendly paint.

- Natural material with low embodied energy.
- Easy to find.
- Comes in a variety of styles, from rustic to contemporary.

- Solid wood doors and worktops can react to changes in temperature and humidity by moving, warping or splitting.
- Worktops can be damaged by water stains and must be periodically oiled or sealed.
- Sustainable timber is still more expensive than standard varieties.

Salvaged wood is strong and full of character.

Bamboo is pale in colour with a distinctive 'striped' look.

Bamboo Units and Worktops

Bamboo kitchen units and work surfaces are now available in a range of styles. Bamboo is a fast-growing grass that can be harvested without disrupting the plant's root system. When grown in a well-managed way, it is a good renewable and sustainable resource.

Method

Existing or new wooden cabinets can be refaced with bamboo; 100-per-cent bamboo units are also available. It is difficult to stain bamboo because it has a high-silica surface that resists colour, but it can be stained or painted after sanding. Ready-finished bamboo units and countertops are widely available.

- Natural and renewable material.
- Distinctive appearance.
- Comparable in cost to sustainable timber.

- Less hard-wearing than wooden surfaces and susceptible to scratches and dents.
- Limited choice of natural colours due to resistance to stain.

Appliances

The choices made when equipping a kitchen will have a large impact on the home's energy consumption. A fan-assisted oven heats up thirty times faster than a conventional oven, so consuming less energy. Gas hobs and ovens use half the energy of traditional electric ones, but induction hobs, which generate heat very quickly from an electric current induced by an electric coil, are the most efficient of all. Some automatically turn off when a pot is removed. Fridge-freezers are responsible for one-third of domestic-appliance energy used in the home. It is important to buy the most energy-efficient model available and to check the seal regularly by shutting the door on a piece of paper: if it falls the seal needs replacing. Washing machines should also be energy-efficient models, and clothes should be washed at the lowest temperature possible – often 30°C (86°F) – to reduce energy consumption. It is better to dry clothes naturally than to use tumble dryers. A fully loaded dishwasher is more water-efficient than washing dishes by hand. An Eco Kettle can be filled to its maximum but will then boil between one and eight cups as required, saving on average 30 per cent of the energy wasted by traditional models.

Method

Energy-saving kitchen appliances are widely available. Labels that provide information on the environmental performance of appliances vary from country to country: Australia, for example, has Energy Rating Labels, and in the UK energy-efficient products have the Energy Saving Recommended logo.

- Energy-efficient appliances cut energy demands and save money.
- Eco appliances are stocked by most major kitchen stores at competitive prices.
- A wide variety of styles and sizes is available for most appliances.

- It is expensive to refit an existing kitchen, but appliances can be replaced in stages, starting with the kettle.

This eco filter kettle boils only a specified amount of water.

Kitchen Planning

Careful planning can save energy in kitchens. Fridges should be positioned away from heat sources, such as cookers and dishwashers, as a fridge next to a heat source takes up to 15 per cent more energy to run. The fridge and freezer should be an appropriate size for the number of occupants: a well-stocked fridge uses less energy than an empty one. Work surfaces should be positioned close to windows to make the best use of natural light. When planning a new kitchen, it is useful to incorporate a large cupboard for a recycling bin.

Kitchens

Highly polished stone suits a variety of kitchen styles.

Stainless steel is increasing in popularity for use in homes.

Paper is particularly suitable for DIY kitchen projects.

Stone Worktops

Natural stone, such as marble, granite, limestone, quartz and slate, provides attractive and hardwearing countertops. The properties of each stone vary: granite, for example, is the most durable and resists chips and scratches, whereas limestone and marble both stain easily, and so require more frequent sealing. Stone surfaces can have a number of finishes, including high-gloss polish, honed – which is smooth but matt in appearance – and tumbled, which is less smooth and looks more rustic.

Method

Before installing stone slabs, it is important to make sure that the kitchen units can support their weight, and to check that the colour is as expected, as it may vary from the showroom sample. If possible, use a locally quarried stone and a VOC-free sealer.

+ • An abundant, reusable natural resource.
  • Extremely hard-wearing.
  • Good variety of types, colours and finishes.
  • Heat-resistant.

– • A non-renewable resource.
  • Quarrying can degrade landscapes, both visually and in terms of biodiversity.
  • Expensive.

Metal Worktops

Metal worktops and units are sleek, durable and hygienic, and recycled stainless-steel and aluminium versions are now available. These are sustainable as well as stylish and contemporary.

Method

Recycled metal can be used in exactly the same way as virgin aluminium and steel. Metal units are usually custom-made and can be designed to incorporate sinks and splashbacks.

+ • Non-porous, limiting growth of bacteria.
  • Resistant to stains.
  • Not damaged by hot pots and pans.
  • Easy to clean.
  • Recycled and recyclable.

– • High embodied energy.
  • Often contain only a percentage of recycled materials, ranging from 50 to 90 per cent.
  • Stainless steel shows scratches and fingerprints, although clear-powder-coated aluminium work surfaces are available and do not suffer from this problem.
  • Expensive, with costs comparable to high-quality granite.

Paper Worktops

It sounds improbable, but kitchen worktops made of recycled or sustainable paper are now available. The paper is treated with non-petroleum, formaldehyde-free resins and then pressed and baked into solid sheets. It comes in thicknesses ranging from 5 to 100 millimetres ($\frac{1}{5}$ to 4 inches) and in a range of colours.

Method

This material is extremely easy to use. Cut and shaped with standard woodworking tools, it is a good choice for DIY kitchen projects. A face mask should be worn during cutting to provide respiratory protection from fine particles. A non-oil-based sealant should be applied annually.

+ • Heat-resistant.
  • Non-toxic.
  • Inexpensive.
  • Easy to work with.
  • Hygienic: impervious and does not support the colonization of bacteria.
  • Allows long spans and extended cantilevers.

– • Can scratch if used as a cutting board.
  • Some manufacturers use petrochemical resin, which is neither biodegradable nor recyclable.
  • Stains if spilt substances are not mopped up quickly.

No two pieces of recycled glass worktop are the same.

Plastic gives a sleek, contemporary appearance.

Belfast sinks combine well with reclaimed timber.

Recycled Glass Worktops

Kitchen worktops are now being made from recycled glass mixed with a concrete or resin base. They come in a range of colours, including multicoloured, depending on the glass used and stains applied. Various sizes of glass chunk can be selected, and some companies will even provide a Certification of Transformation that records the source of the glass.

Method

The green credentials of companies vary so it is important to check what percentage of the finished product is made from recycled glass and what bases have been used. Solvent-free resin is the most environmentally friendly base, although companies that use concrete often use varieties with a high recycled-aggregate content. A representative from the company can install the surface, or it can be delivered cut to size.

➕
- Diverts waste from landfill.
- Strong and durable.
- Heat-resistant.

➖
- Expensive: comparable in price to granite and marble.
- High-energy product compared with natural alternatives.

Recycled Plastic Worktops

Recycled-plastic countertops are made from a wide variety of materials, including yoghurt pots and drinks cups, fused through a combination of heat and pressure so that no binding agents or resins are required. They are available in a range of colours, patterns and textures.

Method

The percentage of recycled content varies depending on the manufacturer and on the buyer's style preference, so the product's constituents should be checked before purchase. Counters can either be installed by the manufacturer or be delivered cut to size. They require very little maintenance and should last at least ten years.

➕
- Diverts waste from landfill.
- Hard, durable and water- and stain-resistant.
- A healthy alternative for those keen to avoid the glues used in such conventional plastic-laminate countertops as Formica.

➖
- High embodied energy, although this can be as low as 50 per cent of the energy required to create sheets from virgin plastic.
- Excessive heat will melt the plastic.

Salvaged Materials

Salvaged materials, such as a reclaimed slate worktop or a second-hand Belfast sink, give a kitchen a rustic character and reduce the materials and energy required to create new products. The possibilities are infinite: even scaffolding boards can be transformed into a worktop and unit fronts.

Method

Architectural salvage yards, many of which list their products online, are a good place to find second-hand kitchen items. Salvaged materials often cannot be incorporated into standardized kitchens, so those wishing to use salvaged items will need to design the kitchen themselves and either commission a carpenter or build their own units.

➕
- Diverts waste from landfill.
- Reduces demand for new products, saving energy and materials.
- Creates a unique kitchen full of character.

➖
- Time and effort are required to find and install these items.
- It can be difficult to create a unified look when relying on salvaged pieces.

30 Sustainable Homes

LOBLOLLY HOUSE

TAYLORS ISLAND
MARYLAND
USA

Left: With its pine stilts and staggered timber cladding, the house merges into the surrounding forest, while making a minimal imprint on the land.

Below: The house was prefabricated entirely from panels with integrated servicing, meaning that it can be quickly and easily disassembled.

The energy consumed during the construction of new buildings and the amount of non-recyclable waste amassed through their eventual demolition are two of the most environmentally damaging processes currently being witnessed across the world. The Loblolly House offers a solution to these problems through prefabrication. As it stands, the building visually disappears into its forested surroundings, but eventually it can be packed away and recycled, leaving the site exactly as it was found.

Stephen Kieran, a founder of Kieran Timberlake Associates, built the house for his wife and their two grown-up children. Located on the edge of Taylors Island in Maryland, the house is surrounded by the loblolly pine trees after which it is named, and overlooks the vast expanse of Chesapeake Bay, the largest estuary in the United States. It is a staggeringly beautiful site, and one that has been preserved through the sensitive design and construction methods of the architect.

Greatly influenced by its forest setting, the building is in some ways reminiscent of a giant tree house, being raised on pine stilts driven 7.5 metres (24½ feet) into the ground, a feature that limited disruption to the site and helps to integrate the building visually with nearby tree trunks. The 186-square-metre (2000-square-foot) house that

The folding glass doors of the west façade give the clients complete control over ventilation while providing views through the loblolly pines to Chesapeake Bay.

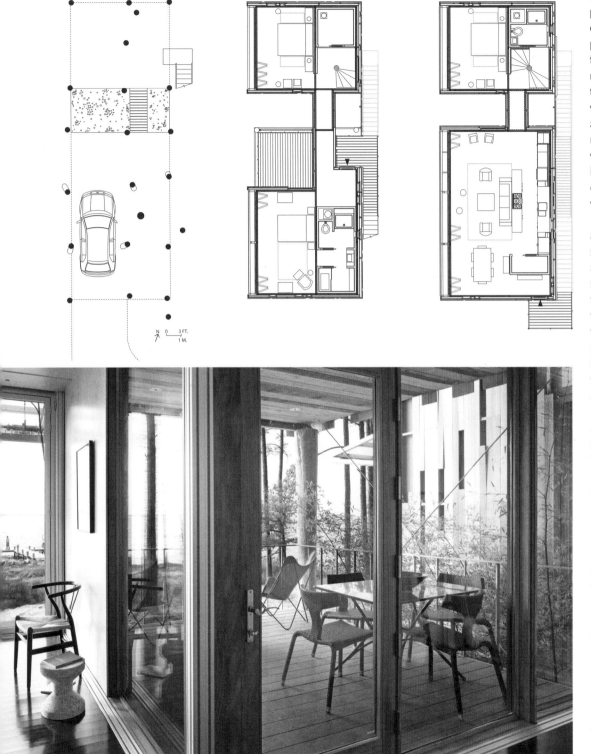

rests on these stilts is composed entirely of prefabricated elements and ready-made components delivered from New Hampshire by lorry. The starting point was an aluminium frame. To this were attached floor and ceiling panels termed 'smart cartridges', made from plywood sheathing and engineered timber and fabricated with built-in electricity cables, water and waste-water pipes, underfloor heating and air ducts. Fully integrated bathroom and mechanical room modules were then lifted into place and the whole enclosed within exterior wall panels containing insulation, windows, interior finishes and exterior cladding. The entire house was assembled on site within six weeks using only a spanner.

The north, east and south sides of the building are wrapped in an abstract timber rain screen of staggered cedar boards laid over glass at some points and solid wall at others, for an effect that evokes the solids and voids of the surrounding forest. The west façade comprises an adjustable glazed system of two layers: folding glass doors on the inside and, on the outside, polycarbonate-clad hangar doors that provide an adjustable awning as well as weather and storm protection. The operable glazed areas allow the occupants to control precisely how much sun and air enters the building at any one time. The three bedrooms, open-plan kitchen and living and dining area all benefit from views through this glass, while service rooms run along the enclosed eastern wall. Water for the house comes from an on-site well.

The architect has been greatly impressed by the efficiency, speed and high-quality finishes of this prefabrication system. Just as the components of this building were assembled quickly and with minimal mechanical help, they can also be disassembled easily and swiftly. The entire building can be deconstructed into its original component parts without leaving any construction debris. This offers numerous opportunities: the house could be re-created on a different site, reassembled in a new configuration, have component parts added or be divided into two smaller dwellings. It is like a giant Meccano set that offers countless possibilities while producing no waste. Kieran believes that such prefabricated methods as these are an effective means by which architects can meet their obligation to design buildings that can be disassembled without creating waste or disturbing their setting.

Opposite, top, left to right: Ground-, first- and second-floor plans.

Opposite, bottom: A sheltered terrace is positioned between the two bedrooms on the first floor.

Above: Interior finishes have been kept minimal, with the aluminium frame left exposed.

Right: A timber staircase at the rear of the property leads to entrances at both levels.

TOTAL FLOOR AREA
186 square metres (2000 square feet)

ARCHITECT
Kieran Timberlake Associates
420 N 20th Street
Philadelphia
Pennsylvania 19130.3828
USA
+1 215 922 6600
timberlake@kierantimberlake.com
kierantimberlake.com

PHOTOGRAPHY
Peter Aaron/Esto/View

ROWE LANE HOUSE

LONDON
UK

Buildings in London are predominantly constructed of brick and concrete. But nestling down a small lane in the heart of gritty Hackney, in the east of the city, is a timber house that challenges prevailing attitudes about how, and with which materials, urban buildings should be constructed. The Rowe Lane House is a timber-framed structure that explores new methods of prefabrication. Built in just sixteen weeks, it not only promotes a new architectural aesthetic but also provides a model for environmentally friendly residential projects that could be applied more broadly to help meet Britain's spiralling housing needs.

Designed by Marcus Lee of FLACQ Architects for himself and his young family, the Rowe Lane House aimed to create a timber system that would ensure fast, efficient and straightforward construction, as well as flexibility once the building was complete. Working closely with the design consulting firm Arup, FLACQ created a bespoke prefabricated, modular timber-frame kit. The structure has also been developed with no internal load-bearing walls, allowing Lee to carve up the interior spaces with demountable partition walls that can be reconfigured as the demands of his family evolve.

The main frame is constructed from glulam (glue-laminated) beams, comprising several layers of timber bonded to produce beams in a variety of shapes that are stronger than solid wood and able to span greater lengths. Because the frame is exposed both inside and outside the building, FLACQ chose glulam made from Siberian larch, a highly durable species, which is grown in sustainably managed forests. The glulam elements are joined together by specially developed stainless-steel connections that are entirely hidden from view. The frame sits on a foundation made of strips of concrete set in 3-metre-deep (10 feet) trenches, resting on the earth's gravel strata – this depth provides sufficient weight to anchor the lightweight house without disrupting the shallow foundations of the adjacent property.

The walls and roof are lined with a 150-millimetre (6-inch) layer of flax insulation – a natural, renewable and non-toxic product – overlaid with Pavatherm insulating fibreboards, a German product that does not contain any glue or wood preservatives. Externally the entire building, including the roof, is clad in cedar, with door and window frames made from the same timber.

The house is arranged so that all the storage and service areas are located at the perimeter, forming a buffer to the party walls and maximizing space within. The ground floor is almost entirely open plan, with large glazed areas visually connecting the interiors to the courtyard in front and the garden behind, and glass doors at either end of the property opening in the summer to cross-ventilate the space. Over the kitchen the roof is fully glazed, drawing natural light into the ground-floor areas. In the summer this glazing is partially shaded by trees on the site, but in the winter, when the trees lose their leaves, the sun helps to heat the home. The bedrooms, on the first floor of the house, have smaller windows for privacy.

Throughout, walls and ceilings are finished with lime plaster and porous paints, both non-toxic

Opposite: Viewed from the rear, this timber house surrounded by trees belies its gritty urban setting.

Left: The house was assembled in just sixteen weeks using prefabricated wooden beams invisibly joined by stainless-steel connections.

Above: The frame of Siberian larch, both durable and sustainable, is left exposed, with a cladding of untreated cedar that will turn silvery grey with age.

An open-plan layout on the ground floor allows natural light to pour in to the kitchen, dining and living areas.

The interior walls are finished with lime plaster and non-toxic porous paints. A wood-burning stove supplements underfloor heating from a wood-pellet boiler.

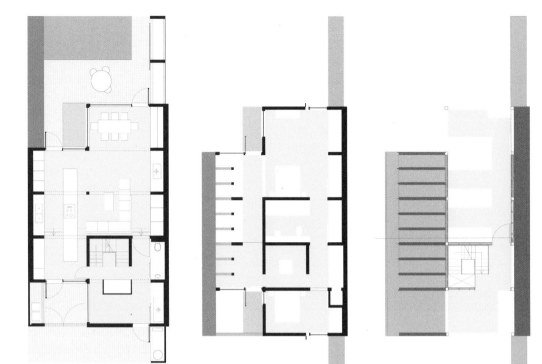

Left, from left: Ground-, first- and second-floor plans.

Below, left and right: None of the internal walls is load-bearing, allowing the interior to be reconfigured as needed, which will increase the longevity of the building.

Opposite, top: The fully glazed roof above the kitchen draws natural light and solar heat energy into the house during the winter, while in the summer it is partially shaded by the surrounding trees.

Opposite, bottom: Storage and service areas are arranged along the sides of the home, maximizing internal space and allowing uninterrupted views through the house to the garden at the rear.

materials that allow the timber walls to breathe, preventing damp and rot. The floors in the living areas are laid with slate, which is easy to clean, hard-wearing and natural. Upstairs, soft carpets have been installed. All the floors are equipped with underfloor heating, fuelled by a wood-pellet boiler. Solar hot-water panels on the south-west-facing roof are used to heat water.

The Rowe Lane House is a modest building that has been carefully planned to provide a comfortable and energy-efficient family home. As Lee's children grow, he can add, remove and alter internal walls to create the most suitable configuration for their needs without additional materials or disruptive building work, extending the life of the building. If the house is ever abandoned it can be dismantled and either rebuilt elsewhere or recycled.

It is rare to see timber buildings in an urban context, and yet the Rowe Lane House slips effortlessly into its surroundings. As the cedar cladding ages it will become even more at one with its garden setting. It is a project that clearly demonstrates the advantages of timber as an efficient, economical and sustainable construction material that challenges the predominance of brick and offers an attractive alternative.

SITE AREA
270 square metres (2905 square feet)

FLOOR AREA
250 square metres (2690 square feet)

ARCHITECT
FLACQ Architects
4 John Prince's Street
London W1G 0JL
UK
+44 (0)20 7495 5755
info@flacq.com
flacq.com

PHOTOGRAPHY
Page 50: Kevin Lake
Pages 51–55: Jefferson Smith/FLACQ Architects/Media10 Images

SAFARI ROOF HOUSE

KUALA LUMPUR
MALAYSIA

The Safari Roof House may not at first seem to be a sustainable building. The 540-square-metre (5800-square-foot) villa in Kuala Lumpur, Malaysia, is essentially a concrete bunker, with floors, walls, stairs, ceilings and bathrooms all featuring the material to a greater or lesser extent. But this house has one innovative feature that proves its ecological worth: its roof. The unusual canopy is designed to insulate and ventilate the interiors naturally, banishing the air conditioning that has become so prevalent in Asia. Combined with the thermal-mass properties of the concrete, it allows the building to regulate its own internal temperature, dramatically cutting the house's energy consumption.

As its name hints, the Safari Roof House's form is based on the Series Land Rovers used in the tropics in the 1970s. A simple sheet of aluminium was held by metal feet off the top of these vehicles, acting as a sun break and sandwiching an insulating layer of air, which, when it reached a certain temperature, simply escaped from the sides of the canopy. Architect Kevin Low of Small Projects adapted the device for this building with a canopy of corrugated bituminous-steel sheet raised from the body of the house on mild-steel hollow sections. Precluding any need for synthetic heat

The key ecological aspect of the design is its canopy roof of corrugated-steel sheet, which acts as a sun break and creates an insulating layer of air to cool the house naturally.

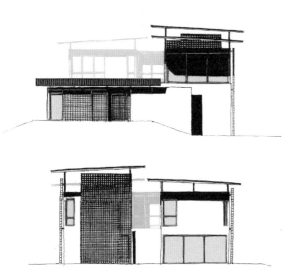

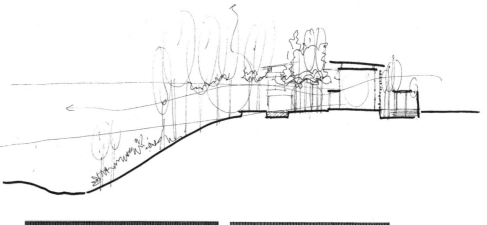

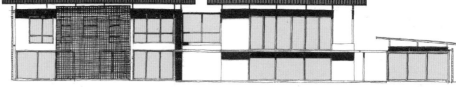

insulation, it is a cost-effective method of using minimal materials to provide natural insulation and ventilation.

The plan of the house is designed around a gravelled garden, planted with thirty-nine trees of eight local species, including *Koompassia excelsa* (Tualang tree), *Caesalpinia ferrea* (Brazilian ironwood) and *Tristania obovata* (Northern box). Selected for their variety of growth rates and heights in order to maximize shade from the late-morning sun, they create a beautiful outdoor garden reminiscent of a forest plantation. The bed of gravel helps limit water run-off during monsoons, and the garden abuts a cast-concrete swimming pool, which marks the boundary of the property.

The compound is entered through a 3-metre-wide (10 feet) steel door, which, when slid shut, seems to cut off the occupants from the outside world. Around the garden three blocks are arranged in a U-shape, with spacious living accommodation on the ground floor and bedrooms above for the young family of four and the grandmother. All rooms have generous glazed walls, many of which slide open on to the courtyard so that indoor and outdoor areas flow into one another.

Above, clockwise from top left: North elevation; section concept; east elevation; south elevation.

Right: The house bends in a U-shape around a gravel garden planted with eight local tree species, their varying growth patterns maximizing shade.

Above: Extensive use of concrete brings high thermal mass and durability to the project. On the west elevation pre-cast concrete ventilation blocks shade the house and provide privacy, while also admitting light.

Right: Generous windows throughout the house connect the rooms with the surrounding landscape as well as providing ventilation.

The house was constructed from a reinforced-concrete frame with a clay-brick infill and cement render finish. On the west elevation, large walls of pre-cast concrete ventilation blocks shade the house and provide privacy. Usually used in a single layer beneath ceiling level, the utilitarian blocks are here stacked to form entire walls, creating interesting patterns inside the house where the sun seeps through. Operable doors and windows in the inner glass shell that backs these walls can be opened to allow air to flow through the ventilation blocks and into the house. The thermally massive walls have proved an effective tool in keeping the house constantly cool without mechanical assistance.

The interior finishes are all variations on the common material of concrete, from polished floors and rendered walls to ceilings that express the marks of the plywood shutters in which they were cast. The house appears as a concrete shell, with as few additional materials as possible; there are no fancy wallpapers or expensive stone floors. The fabric of the building is fully exposed, and the result is surprisingly sophisticated.

It is indisputable that concrete is energy-intensive to manufacture, but here the architect has played on its advantages to create a thermally efficient building that can regulate its own internal temperature even in the extremes of the Malaysian climate. Concrete is also a strong and durable material, and this house should stand for a long time with little maintenance. Whether it becomes coated with moss, overrun by creepers or stained by dripping water, the exterior of the building will no doubt evolve as the structure slowly but surely pays back its carbon credit.

Below: Natural light seeps in through the concrete ventilation blocks, and when the inner glass panels are opened a cooling cross-breeze blows through the blocks into the living space and out to the courtyard beyond. Generous windows throughout the house allow the rooms to connect with the surrounding landscape.

Opposite: The continuation of the polished concrete floor from interior to exterior helps create a visual connection between the two spaces. Shutter marks on the ceiling of the covered walkway express the process by which the concrete was set.

TOTAL FLOOR AREA
540 square metres (5800 square feet)

ARCHITECT
Small Projects
#1 Jalan Tenggiri
Kuala Lumpur 59100
Malaysia
+6012 200 1800
isd@pd.jaring.my
small-projects.com

PHOTOGRAPHY
Pages 56, 58b, 59b, 60–61: Olivier Ike/Ike Branco
Page 59t: Kevin Low

MAISON A/STUDIO B

PARIS
FRANCE

This collaborative project between three French architects has seen a run-down Parisian house transformed into a high-tech and impressively sustainable home. The project is divided into two adjoining properties: Maison A, the family home of the clients; and Studio B, a smaller house, which they let. The clients asked the architects to open the house to the garden and expand the existing floor area to create light and airy interiors with open but clearly defined spaces that encourage interaction between family members. They also wanted the houses to be energy-efficient and hence economical to run, with reduced carbon emissions. The result is a dramatic transformation that combines sensitive renovation techniques with cutting-edge materials and gadgetry. While the street façade has retained its original character, the back of the property has been entirely reconstructed with extensive glazed areas and futuristic metal façades. Enclosed within this shell are two exciting, contemporary and environmentally friendly homes.

Built in 1860, the original house was small, not particularly attractive and configured so that nearly all the rooms looked on to the road. However, it did have a garden, which is rare in Paris, and it backed on to a picturesque and quiet little passageway. The client asked architect Jacques Moussafir to turn the house into a family home, and when a studio overlooking the garden came up for sale, employed

The front of the house retains its nineteenth-century character, but the rear has been transformed to open the house to natural light and views of the garden.

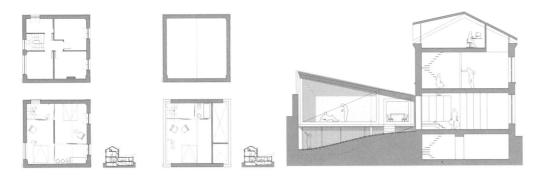

Isabelle Denoyel and Eric Wuilmot to transform that into a contemporary apartment.

The interiors of both properties were reconfigured to create more flowing and practical spaces. The shell of Studio B remained the same but the ground floor was carved into a kitchen and living-room. Two mezzanine levels above are connected by a staircase and stacked one on top of the other. A bathroom and study are on the first floor, and the space under the pitched roof was converted into a bedroom.

Moussafir's clever alterations to Maison A transformed it from two to four storeys, almost doubling the floor area to 225 square metres (2420 square feet). This was achieved by converting the cellar and by flicking the roof up at the back to allow comfortable spaces to be created within the loft. An extension at the back of the house was almost entirely rebuilt, creating further rooms at both basement and ground-floor levels. Maison A now contains generous bedrooms for the parents and their three children, as well as four bathrooms, a large kitchen area and living-room, a laundry, wine cellar and multimedia room. Almost all these rooms have large glazed walls looking out over the garden.

The materials used are environmentally friendly. The timber – softwood for the structural frames, with exotic wood used only for the louvres on the extension – comes from sustainable forests. The

windows have low-emissivity, double-glazed panes fitted with filters to cut glare. The polyurethane panels that line the walls are twice as effective per thickness as standard glass-wool insulation. The roof of Maison A has rock wool insulation and is clad in zinc, a totally recyclable material that is not only extremely durable and easy to maintain but also commonly used on Parisian roofs, helping to contextualize the building. Six solar hot-water panels, discreetly mounted flush within the zinc surface, heat enough water for the kitchen and all the bathrooms. Water run-off from the roof, channelled through a drainage system into two 750-litre (165-gallon) storage tanks, is used in the irrigation system that has been installed in the garden.

While in Studio B heating is provided by an energy-efficient condensing boiler, Maison A is equipped with a combination of systems that allows the clients to dictate the temperature in each room: underfloor heating at basement and ground-floor level; radiators powered by a condensing boiler on the first and second floors; and a wood-burning stove with adjustable settings in the main living-room. A double-flux Canadian ventilation system cools and heats the home while introducing fresh air into the interiors. Two metal ventilation shafts have been planted at opposite sides of the garden. One draws in fresh air, which passes underground and into the building, while the other draws out stale air and releases it into the atmosphere. At least half the air volume of each room is replaced every hour. Incoming air passes though pipes buried 1 metre (3 feet) underground, where the earth's temperature varies from 12°C to 15°C (53.5–59°F), and the air adapts to the ground's temperature by thermal exchange. Central controls automatically switch on the system when external temperatures are below 12°C (53.5°F) or higher than 27°C (80.5°F), limiting the client's heating needs and replacing an air-conditioning unit.

Numerous applications within both Maison A and Studio B further reduce their energy and water demands while creating a fresh and toxin-free environment. The walls, for example, are coated with paints low in VOCs, and light fittings are a mixture of LED and compact fluorescent bulbs. Appliances in the kitchens have high energy-efficiency ratings, and the bathrooms are fitted with dual-flush toilets. For every decision that needed to be made, the clients

and architects found an environmentally friendly solution. Both buildings are 40 per cent more efficient than standard new-build houses in France and 20 per cent more efficient than is required by the French thermal regulations of 2005. These architecturally daring and contextual renovations prove that new sustainable technology can be cleverly concealed within chic and practical homes.

Opposite and left: The kitchen in Maison A, positioned in the old part of the building with the original window overlooking the street, opens out on to the living area, which occupies the entire ground floor of the new extension.

Below, left: The first-floor mezzanine in Studio B contains a bathroom and study; the bedroom is positioned above.

Below, right: Two mezzanine levels are stacked above the ground-floor kitchen and living area in Studio B.

TOTAL FLOOR AREA
Maison A: 225 square metres (2420 square feet)

ARCHITECTS
Maison A: Moussafir Architectes
5 rue d'Hauteville
75010 Paris
France
+33 (0)1 48 24 38 30
moussafir.archi@wanadoo.fr
Studio B: Isabelle Denoyel and Eric Wuilmot

PHOTOGRAPHY
Paul Kozlowski

CAPE SCHANCK HOUSE

VICTORIA
AUSTRALIA

When one first catches sight of the Cape Schanck House it seems incongruous. Planted among the modest holiday homes of the Mornington Peninsula, 88 kilometres (55 miles) south of Melbourne, the house has a futuristic and slightly forbidding feel, more spaceship than family holiday home. Yet the design of the house was inspired entirely by its surroundings, establishing a dynamic prototype for a sustainable house that can cope with Australia's increasingly harsh climate.

Paul Morgan built the house with his sister in 2006 to provide a weekend retreat for both their families. The 160-hectare (395-acre) site is in an exposed spot on the southern side of the peninsula, with strong prevailing winds. But Morgan and his sister were attracted by the abundance of native tea trees that grow there, twisting their trunks at dramatic angles to reach the sunlight. They decided to position the house so that the vegetation would remain undisturbed, locating it near the western boundary of the site, where the trees form a natural canopy.

The shape of the 175-square-metre (1880-square-foot) building was developed by analysing the wind and sun conditions using computer renderings and cardboard models. This produced the aerodynamic shape of the sleeping wing, which is clad in Ecoply. Vertical louvres attached to the bedrooms can be adjusted to block out the summer sun, and wind scoops on the façade shade the bedroom windows while capturing cooling winds to cross-ventilate the house naturally. Within, three bedrooms of equal size lead off a central corridor. Morgan deliberately omitted a master bedroom to reflect the communal nature of the house shared by the two families. He also provided two bathrooms to avoid queues. The floors and ceilings of the bedrooms follow the geometric planes of the exterior walls and are lined with exposed plywood.

Jutting out from the centre of the sleeping wing is the main living area, its monolithic external skin of Alucobond wrapping over the roof and digging its edges into the ground. It is in this main living area, the hub of the house, that Morgan's innovative thinking comes to the fore. A huge, bulbous water tank drops down from the ceiling into the centre of the room, exploding all conventions. It is both the architectural and the conceptual focal point of the house, symbolically taking the place of the hearth.

Structurally, it carries the load of the roof and divides the space into four zones: kitchen, living, dining and work. But it is the environmental benefits that are truly revolutionary. The water tank collects and stores rainwater harvested from the roof through a concealed box gutter, the 6-millimetre ($^1/_4$-inch) mild-steel walls of the tank keeping this water at a temperature close to 21°C (70°F). This ambiently cools the house's interior and negates the need for air conditioning. Any excess water drains to an external tank and is used for flushing the toilets, irrigating the garden and washing wetsuits.

While the street façade has guarded slits for windows, at the rear of the property the Alucobond gives way to floor-to-ceiling sliding glass doors, leading to a small terrace – the modern equivalent of the Australian veranda – which basks in the

Opposite: A dramatically angled timber-decked ramp leads to the main entrance of the house.

Below: The house has been positioned to preserve the native tea trees on the site.

Left: Glass doors at the back of the property open on to an external terrace, extending the living area into the garden.

Above: The main living area is wrapped in a skin of Alucobond and large panes of glass.

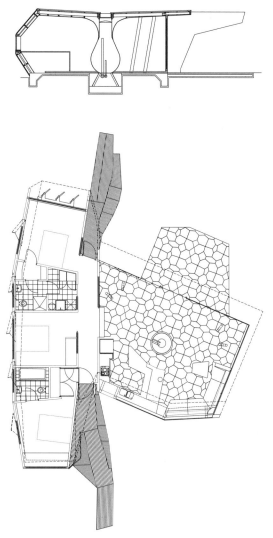

Above: Section (top) and site plan.

Right, top: Custom-made paving stones in the living area derive their shape from nearby rock formations created by fast-cooling lava.

Right, bottom: The bulbous water tank that drops down from the ceiling into the centre of the living area stores rainwater, which also serves to cool the interiors.

afternoon sun. The terrace's custom-made paving stones, their pentagonal and hexagonal shapes derived from nearby rock formations created by fast-cooling lava, continue into the living area, blurring the boundary between inside and out. Like the plywood of the bedrooms, they add an earthy feel, bringing warmth to an interior that is predominantly painted a pristine white.

Australia has a famously arid climate that is becoming even more drought-prone as a result of global warming. Water is the country's most precious commodity, and water tanks are being widely acknowledged as the sustainable answer to house-hold demands. As with many environmentally friendly solutions currently being developed – solar panels and wind turbines included – water tanks have traditionally been stuck on to a building after it has been completed. They are highly visible add-ons, which in many cases damage the architectural integrity of the building. By not only integrating the water tank into the house but also making it a key sculptural element, Morgan has opened a new chapter in sustainable architecture, proving that with imaginative thinking, eco-friendly solutions can be central to the design of a building and enhance, rather than detract from, its appearance.

Left: Exposed plywood surfaces soften the angular planes of the bedrooms.

Below: Viewed from the driveway, the two separate wings of the house, one clad in Ecoply and the other in Alucobond, create a striking composition.

SITE AREA
160 hectares (395 acres)

TOTAL FLOOR AREA
175 square metres (1880 square feet)

ARCHITECT
Paul Morgan Architects
Level 10, 221 Queen Street
Melbourne
Victoria 3000
Australia
+61 (0)3 9600 3253
office@paulmorganarchitects.com
paulmorganarchitects.com

PHOTOGRAPHY
Peter Bennetts

TORQUE HOUSE

HEYRI ART VALLEY
GYEONGGI-DO
SOUTH KOREA

The Torque House is located in Heyri Art Valley, an arts community established by writers, film producers, architects, musicians and artists in Gyeonggi province, South Korea. There, artists can live and work in one place, making, exhibiting and selling their creations. It is a cultural hub that currently has more than forty museums, galleries, concert halls and bookshops and that will continue to grow until the 400 lots in its masterplan are filled.

The clients of Torque House are typical Heyri Art Valley settlers: a still-life painter and a classical-music recording engineer. The challenge facing Seoul-based architects Mass Studies was to combine these activities in one home, while creating a house that was sympathetic to its mountainous surroundings. The solution is a dynamically shaped building with unusual moss-covered walls.

The clients had very particular demands when it came to the interior. The painter asked that her studio be visually closed to the outside so that there would be no distractions, but wanted it to benefit from natural light. She wanted a double-height space that was rectangular to provide a sense of stability, connected to the rest of the house yet with a separate entrance. The recording engineer asked for a music room that could accommodate up to thirty orchestral performers, connected by a soundproof window to a control room where he could monitor, record and edit music. It needed a separate entrance and circulation route to the rest of the house, so he could avoid disturbing his wife, and he asked for a small kitchen and outdoor space where visitors could relax during recording sessions. The acoustics of the music room were the crucial element of the brief as well as

the most demanding for the architects, dictating every aspect of the design, from the irregular shape of the walls to the details of the windows. The house also needed to incorporate comfortable living and sleeping areas for the couple and their two children.

The result is a 441-square-metre (4750-square-foot) building rising three storeys as one solid mass. Its slanted shape reflects the curved boundary of the site, producing a twisted rectangular form that inspired the name of the house. Inside, the ground floor is split in two, with half reserved for the double-height artist's studio and the other providing the family's kitchen, dining and living area, with three bedrooms and a family bathroom above. The second floor contains the music studio, a rest area for musicians and a small kitchen. At the top is a sculptural roof terrace.

Opposite: The twisted rectangle of the Torque House rises dramatically from its wooded setting.

Left: The south-west (top) and north-east sides are clad with turf mats planted with moss, chosen for its environmental and aesthetic benefits. The south-east façade is entirely covered by semi-reflective glass, which brings light into the building.

Above: The richly textured moss contrasts with the smooth glass of the long strip windows.

On the shorter, north-west façade, the house's concrete frame – a construction technique that is ubiquitous in Korea – has been left exposed, but the south-east façade is clad entirely in semi-reflective glass. This provides views out from the residential areas on the ground and first floors while protecting the privacy of the occupants. It allows natural light to enter the building but prevents rapid heat gain and glare in the summer months. The thermally massive walls gradually absorb and release radiant heat, passively regulating internal temperatures.

From both an aesthetic and a sustainable perspective it is the north-east and south-west elevations of the building that are the most exciting. Each has a

Top: Both clients have been provided with generous workspaces: a studio for the artist on the ground floor (left) and a music room for the sound engineer on the second floor (right).

Centre and bottom: Sections (above); south-east elevation (below left) and first-floor plan.

Opposite: The exterior spaces are as dramatic as the interiors. A semi-open stair leads to the sculptural roof terrace.

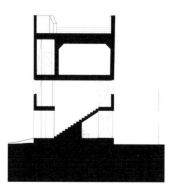

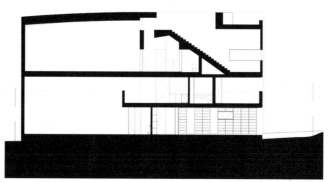

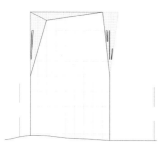

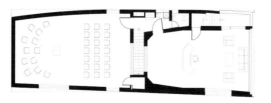

long strip window at the centre with further smaller windows scattered across the surface, which consists of an innovative Japanese product called the Moss Catch System. Never before used to clad a building in Korea, the system comprises turf mats planted with two types of moss, which grow well regardless of their orientation, and so are suitable for both sides of the building. The architects chose the material to add visual interest to the imposing three-storey walls and because of its many environmental benefits. As with a green roof, the moss coating prevents water run-off, protects the house from wear and tear and provides excellent acoustic and thermal insulation. Here, the system also helps to control the internal temperature of the building, as water pipes installed behind the cladding to irrigate the moss cool the building in the summer. Since moving in, the clients have hardly used the air conditioning, and have saved an enormous amount of energy.

By using such adventurous materials as moss and semi-reflective glass, Mass Studies have created an eye-catching and energy-efficient building that meets an exacting brief. It is a thoughtful addition to a community bursting with creativity.

SITE AREA
534 square metres (5748 square feet)

TOTAL FLOOR AREA
441 square metres (4750 square feet)

ARCHITECT
Mass Studies Architects
Fuji Building 4F
683-140 Hannam 2-dong Yongsan-gu
Seoul 140-892
South Korea
+82 (0)2 790 6528/9
office@massstudies.com
massstudies.com

PHOTOGRAPHY
Kim Yong Kwan

SOLAR UMBRELLA

VENICE
CALIFORNIA
USA

The architectural heritage of Venice, California, is as bohemian as its inhabitants. Modern houses by such architects as Frank Gehry and David Hertz are scattered among rows of colourful bungalows, while the distinctive arched architecture commissioned by Abbot Kinney, who developed Venice as a resort and amusement park in the 1890s, still provides a backdrop for the muscle men on Venice Beach. It is within this architectural melange that Angela Brooks and Lawrence Scarpa have created Solar Umbrella, an unashamedly modern house that is designed entirely around the concept of sustainability.

Brooks and Scarpa are dedicated and vocal supporters of sustainable architecture. They co-founded Livable Places, a not-for-profit organization that raises public support for sustainable design, and are both principals at Pugh + Scarpa, the architectural firm behind such buildings as Colorado Court, a solar-panel-clad apartment block in Santa Monica, named a finalist in the World Habitat Awards of 2003.

The transformation of their 1920s Spanish-style home from a 60-square-metre (645-square-foot) bungalow to a 177-square-metre (1900-square-foot) double-storey house was inspired by Paul Rudolph's Umbrella House, built in Sarasota, Florida, in 1953. Designed with a wooden trellis arching over its roof, swimming pool and terrace, the Umbrella House was revolutionary in its time for attempting to mitigate the extreme summer sun through its external envelope. Scarpa and Brooks have taken this idea and developed it even further, wrapping around their house a protective arm lined with solar panels. This is not only a shield against the sun but also an integrated architectural feature that harnesses the sun's power and converts it into electricity. It is a fairly simple concept, but it has been executed with such architectural flair that the panels become just one of many visual layers contributing to the complex montage of the building.

The Solar Umbrella house sits on a small site (430 square metres/4630 square feet) jammed in between two residential streets. In order to optimize exposure to the sun, the architects reversed the orientation of the original building. What was previously the back garden is now an impressive entrance courtyard complete with cast-concrete swimming pool and fishpond, which must be crossed on stepping stones to reach the front door. The carcass of the original bungalow now houses the kitchen and dining areas, a study, and a bathroom and bedroom for the couple's son. To this the architects added a new entrance, a utility room, an impressive living area connected to the front courtyard by a great sliding pane of glass, and a first-floor master bedroom and bathroom. From outside, the new configuration at first appears confusing but is clear once it becomes apparent

Left: A canopy of solar panels wraps around the house, both providing shade and harnessing the sun's energy to supply 95 per cent of the building's power.

Opposite: Exterior flows freely through to interior to bring in natural light and air, with a large outdoor deck over most of the first floor and a vast sliding door opening the living area to the front court.

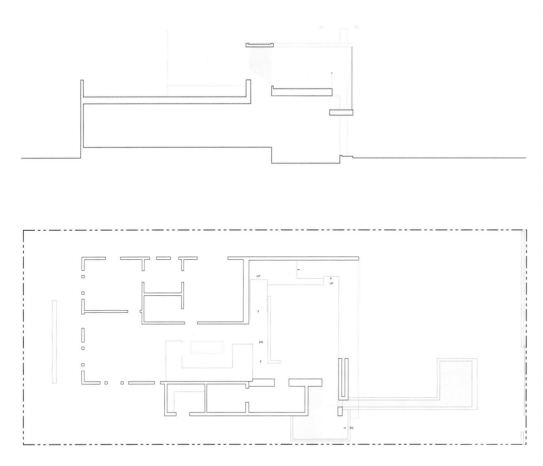

Opposite: A large steel drain along the front of the house collects rainwater run-off, which is held in an underground basin.

Above, from top: Longitudinal section; first-floor plan.

Right: The original house has been incorporated into the new building with the minimum of demolition, and reorientated so that the old front entrance now constitutes the back.

that much of the first floor is an exterior deck leading off the master bedroom.

Every effort was made to conserve building resources and reduce waste. Only the southern wall of the original bungalow was pulled down to link it to the extension. Most of the wall framing and all the roof and floor framing of the old building was incorporated in the new structure, and the original stucco was even sandblasted and skim-coated with a new colour to match the addition. Since concrete manufacture produces high levels of carbon dioxide, the concrete used to construct the extension, cast between shutters made from reclaimed timber, was adulterated with 50 per cent fly ash. Recycled mild steel, rusted and then sealed, was used for the exterior supports, and the stucco that coats the extension has an integral coloured pigment so that painting will never be required. Around 85 per cent of the construction debris – mostly from a garage that was pulled down and replaced with a carport equipped with plug-ins for the owners' electric car – was sorted and sent for recycling.

The house was designed to regulate internal temperatures with minimal mechanical assistance. The concrete floor and walls are strategically placed to act as heat sinks, helping both to cool and to heat the building. Insulation made from recycled paper pumped into both the new and the existing walls boosts the thermal value of the walls to 75 per cent above that of a conventional timber-frame building. Underfloor heating was installed throughout the house, and all glass surfaces are double-glazed, krypton-filled, low-emissivity panes.

Most of the glazing is on the south side of the building, where it is partially shaded by the solar-panel canopy and a series of abstract fins that block out the high summer sun, while admitting lower rays to help heat the building during the winter. Operable windows and doors, including skylights in the kitchen and bathroom, are positioned to encourage natural cross-ventilation, and the perforated steel staircase allows hot air to rise through and out of the building. As a result, 92 per cent of the building is naturally ventilated and the whole is naturally lit during daylight hours. These design strategies, combined with the use of energy-efficient appliances, mean that the eighty-nine solar panels can provide 95 per cent of the house's energy needs, exceeding by 50 per cent

both the state of California's Title 24 Energy Code and the city of Santa Monica's Green Building Design and Construction Guidelines.

Water is also carefully managed. Rainwater is collected from the roof by a large steel drain, visible at the front of the building, and flows directly into an underground retention basin. This allows most of the water to remain on site and prevents water run-off from collecting oil, rubbish and pesticides in the streets before it reaches the nearby sea. Both the pool and the domestic hot-water supply are heated by solar hot-water panels fixed to the roof. The washing machine and dishwasher both use less water than traditional models; the kitchen tap and showerheads are low-flow fixtures; and the existing toilet was replaced with a dual-flush model free of charge through a city rebate programme.

Cost-effective and eco-friendly materials are used in unconventional ways. Homasote acoustic panels were sanded and used as a finish material for the built-in furniture, and the floors are made from exposed polished concrete and orientated strand board (OSB). The sofa is built in, with storage space underneath, to maximize the living space.

Solar Umbrella is an exemplar of green design. Its eco-friendly technology will pay for itself in ten years through energy and water savings. Rather than just another address on the architectural tour of Venice, this house offers a cure for California's energy-guzzling habits, proving that ecologically sound architecture can compete both financially and aesthetically with the most revered of California's Modernist buildings.

Above: Interior fittings were selected for their ecological performance: the kitchen floor is laid with orientated strand board (OSB) panels, made from leftover woodchips, and the units are of Homasote, which is made from recycled newspaper.

Opposite, top: The ground floor is open-plan, with views stretching from the living-room through to the kitchen and dining area. A perforated steel staircase allows hot air to rise within the building.

Opposite, bottom: The first-floor bedroom has extensive built-in storage and an en suite bathroom.

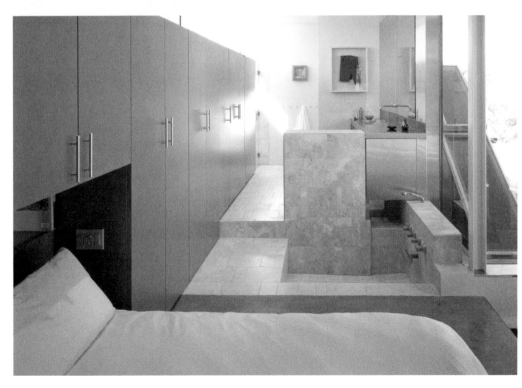

SITE AREA
430 square metres (4630 square feet)

TOTAL FLOOR AREA
177 square metres (1900 square feet)

ARCHITECT
Pugh + Scarpa Architects
2525 Michigan Avenue, Building F1
Santa Monica
California 90404
USA
+1 310 828 0226
info@pugh-scarpa.com
pugh-scarpa.com

PHOTOGRAPHY
Marvin Rand, courtesy of Pugh + Scarpa Architects

WALLA WOMBA GUEST HOUSE

BRUNY ISLAND
TASMANIA
AUSTRALIA

Walla Womba Guest House is a self-sufficient building immersed in the coastal bushland of Tasmania. Its green credentials were partly born of necessity, as the remote site has no municipal water, power or sewer connections. But 1+2 Architecture has also shown a deep respect for the surrounding landscape, which is rich in native flora and fauna. With little visual or environmental impact, they have created a contemporary house that harnesses the earth's natural resources and slips politely into its rugged surroundings. It is an idyllic retreat with impressive sustainable principles.

The 250-square-metre (2690-square-foot) house sits on an 80-hectare (200-acre) site bordering Tasmania's Tinpot Bay, its precise location selected for coastal views and to disrupt the site as little as possible. The two parallel pavilions, connected by a central circulation space, are raised on a steel-frame subfloor, which minimized the need for excavation and preserved the ground underneath. Aesthetically it gives the house a sense of weightlessness, while symbolically it signifies the architects' quest to have as little impact as possible on the surroundings.

The clients, an American couple who live a not-so-green thirty-hour plane journey away in Rochester, New York, had initially asked for a two-

Below: Raised off the ground but sitting lower than the surrounding trees, the house makes little physical or visual impact on its bushland setting.

Opposite: The overhang of the roof was calculated to allow solar gain in the winter but prevent excess heat entering the home in the summer.

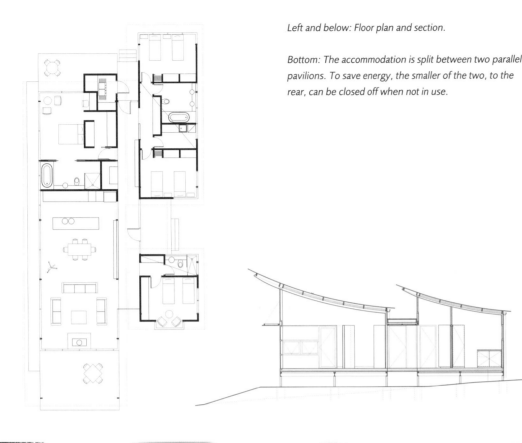

Left and below: Floor plan and section.

Bottom: The accommodation is split between two parallel pavilions. To save energy, the smaller of the two, to the rear, can be closed off when not in use.

storey building, but the architects convinced them that a one-storey structure would reduce the roof's visual impact by keeping it below the level of the trees. The building's steel-and-timber frame is clad in oiled timber and dark-grey sheets of steel, chosen to merge with the colours of the surrounding bush. Large double-glazed windows look out on to the coast, visible beneath the surrounding tree canopy. They were carefully positioned to maximize solar gain and to cross-ventilate the interiors naturally, and most open fully, eliminating the boundary between indoors and outdoors. On the north elevation the overhang of the roof was precisely calculated using a combination of computer technology and on-site observation to allow as much low winter sun into the building as possible while shielding the interior from the more extreme summer sun.

The interior spaces are divided into distinct public and private areas, with three bedrooms in the smaller of the two pavilions, and the master bedroom and living space under the main roof. A sliding door separates the two pavilions, effectively turning the house into two separate properties, a handy tool when noisy guests come to stay but also a practical way of reducing heating demands when only part of the house is occupied. Decorative finishes of simple hardwood floors, neutral carpets and off-white plasterboard allow the views to dominate.

Such appliances as a dishwasher, a clothes dryer and even a hot tub are testament to how well the architects have designed the building to take full advantage of natural energy and water supplies. Most of the house's power comes from six photovoltaic panels mounted on a disused shipping container parked in a sunny clearing 30 metres (100 feet) from the house (although portable gas cylinders fuel the hot-water system and cooking appliances). Rainwater is ingeniously syphoned from the roof through holes punched in the valleys of the corrugated-steel sheets that make up its swooping form. Underneath these holes, which are too small for leaves to slip through, the water collects in a concealed gutter and is piped down the side of the house to two underground 12,000-litre (2600-gallon) tanks, from where it is channelled to the kitchen and bathroom. A similarly discreet and effective system handles household waste, with bathroom and kitchen waste collected underground and processed in a septic tank before being dispersed around the garden via a

network of subterranean trenches. The waste evaporates or composts into the earth, where its nutrients are absorbed by native shrubs planted along the trenches. Combined with high levels of insulation, these cleverly integrated systems allow the house to be almost self-sufficient without having to rely on unsightly clip-on technology.

There can be no disguising the fact that holiday homes are inherently unsustainable buildings, especially when the clients live on a different continent. But architects are to a certain extent at the mercy of the clients who commission them. By creating a building that so effectively minimizes its impact on its surroundings, 1+2 Architecture has dealt admirably with the challenge.

Fully glazed walls in the open-plan living areas provide views through the trees to the coast beyond, as well as drawing in air and natural light.

SITE AREA
80 hectares (200 acres)

TOTAL FLOOR AREA
250 square metres (2690 square feet)

ARCHITECT
1+2 Architecture
31 Melville Street
Hobart
Tasmania 7000
Australia
+61 (0)3 6234 8122
mail@1plus2architecture.com
1plus2architecture.com

PHOTOGRAPHY
Peter Hyatt

VILLA ASTRID

HOVÅS
GOTHENBURG
SWEDEN

Wingårdhs Architects faced an unusual challenge when they were commissioned to design Villa Astrid. Tucked between two existing houses, the site consisted almost entirely of a steep, rugged rock face. In addition, there were rigid restrictions on the height and angle of the roof. These were not ideal conditions in which to create a two-storey family home. Yet the architects responded with a building that embraces the difficult terrain. Its dark, brooding form extends from the rock face, with the boulders at one point breaking through the skin of the house and invading the private domain. Bold and daring, it takes a hostile section of land and sensitively tames it while creating an energy-efficient, light and spacious home.

The clients are two busy doctors who wanted a manageable and economical house for their family of four. In order to create a two-storey building without violating planning stipulations, the architects dug deep into the site, creating a sunken court enclosed on one side by the rock face. Around this external decked area, the house is bent into a distorted U-shape, the north-facing arm angled to take full advantage of views across the sea, while the eastern wing backs up against the drive and road, and the stunted southern arm provides a carport and storage room.

The house is constructed almost entirely of concrete, chosen by the architects because its high thermal mass will help to keep the house warm in winter and cool in summer. Solid lightweight concrete plastered inside and out has been used for the walls, while the roof is of concrete cast *in situ*, with 400 millimetres (15³/₄ inches) of Foamglas insulation. Covering the whole is a monolithic skin of black, pre-patinated copper sheets that will slowly oxidize to a green verdigris, like lichen on a rock face, obviating the need for maintenance. Limestone gravel around the base of the building neutralizes toxic copper ions released from the metal.

Cutting into this metallic coat, vast angular windows, triple-glazed and set in pine frames, have been strategically placed to optimize passive solar gain, helping to heat the concrete shell. At the end of the north-facing arm, a gigantic rectangular window slices into a seam-drilled slit in the rock, which appears to continue unobstructed into the building, the break preventing moisture and the outside temperature of the boulder from passing through into the house. Around the courtyard, the exterior walls feature double-height glazing, while windows on the north-facing façade exploit views and allow cross-ventilation in the summer.

The 400-square-metre (4300-square-foot) interior is divided into public and private spaces. On the ground floor are the two children's bedrooms, the parents' bedroom, two bathrooms and a cosy

Left: The house makes a virtue of its sloping, rocky site, its distorted U-shaped form nestling into the rock face.

Opposite: The north façade looks out to sea. Extensive use of concrete provides temperature-regulating thermal mass, and the skin of copper will weather naturally to a green colour, requiring little maintenance.

The rock enters the building at the end of the northern arm, a vast window cutting through its mass.

living area, all with sliding doors leading out on to the enclosed deck area. Upstairs is an open-plan kitchen, a separate dining-room and a large living area with sliding doors opening on to a north-facing deck with steps down to the sea. At the far end of the living area, a mezzanine level is suspended above the internal rock face, providing a dramatic addition to the interior and a practical study area for the adults. Finished with ash flooring and built-in furniture, the interiors are surprisingly light given the somewhat menacing external appearance of the building.

The outdoor terrace areas are decked with ThermoWood, for sustainability and durability. Hot-water pipes have been laid under all the floors in the house, concrete being particularly well suited to underfloor heating. These are fed by a ground-source heat pump, which extracts latent heat from the ground to provide the house's hot water.

Wingårdhs say that this is the most technically complex project their office has yet completed, and they are understandably proud of the results. Burrowing into the rock face, Villa Astrid combines the physical robustness of a cave with bright and contemporary internal spaces that capitalize fully on the view. It consumes minimal energy, requires negligible maintenance and, like the rock face from which it emerges, will last a long time.

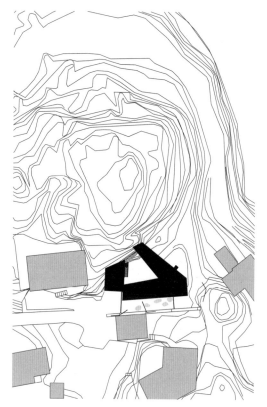

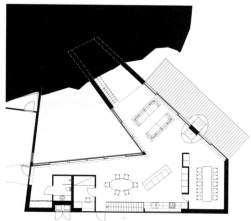

Above, clockwise from left: Site plan; section and ground-floor plan.

Right: The rock face against which the house is built is made a feature by being brought into the living area. A mezzanine study area is suspended above it.

Opposite: Large triple-glazed windows capture the heat of the sun to warm the house, and sliding glass doors around the sunken courtyard, and to the north, bring in light and air.

TOTAL FLOOR AREA
400 square metres (4300 square feet)

ARCHITECT
Wingårdhs
Katarinavägen 17
SE 116 45 Stockholm
Sweden
+46 (0)8 447 40 80
wingardhs@wingardhs.se
wingardhs.se

PHOTOGRAPHY
Pages 88–89, 92b, 93: James Silverman
Pages 90–91: Gert Wingårdh

AVIS RANCH RESIDENCE

CLYDE PARK
MONTANA
USA

Montana is deeply divided over complex problems of development. The state attracts and needs new occupants but grieves over the loss of its rural character as aggressive architectural intrusions are made on its fragile and largely pristine landscape. Most newcomers construct large, modern buildings on the ridge tops, far from the roads. Historical rural buildings are often left to fall into disrepair or, if they are renovated, are frequently sold to become tourist attractions in new locations. With their design for the Avis granary and house, however, Fernau & Hartman Architects have shown extreme sensitivity to both Montana's history and the local environment.

The Californian client, who had combined several properties to create a 7000-hectare (17,000-acre) site, asked the architects to design not only a comfortable home for his family and their guests but also the ancillary buildings necessary for a working ranch. One of the first questions was where to position the residence. After a year studying the land, consulting neighbours and carrying out environmental studies, the architects identified six possible locations. A question posed by the client – 'What would the farmer do?' – pinpointed the final site: the farmer would build on the rural road that bordered the property, providing easy access to the nearby community of Clyde Park. In fact, a number of vacant buildings already lay in a neglected state in close proximity to the road. Swallowing their grander plans for a new design, the architects instead expressed

Left: The accommodation is arranged in two groups: the granary cluster and, seen in the background, the barn cluster. Both revive neglected historical buildings.

Below: The granary cluster comprises the family house, which is set in a former farmhouse (left), guest rooms and basketball court set in an old granary (centre), and a carport and workshop (right).

Bottom: The barn cluster is the working part of the ranch and houses (from left to right) a stable, office, haybarn and guest quarters.

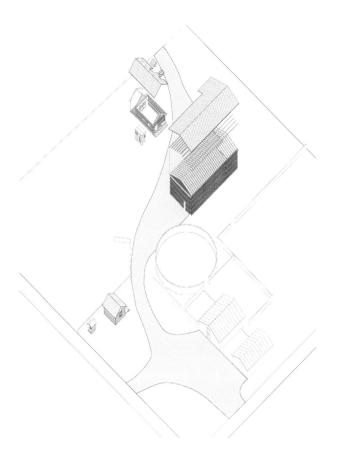

Above: Site plans of the granary cluster (left) and the barn cluster (right).

Right: In the barn cluster, a small two-cell granary has been converted into guest accommodation.

Opposite, top: The new haybarn is contextually sensitive, using traditional timber building techniques yet avoiding pastiche.

Opposite, bottom: From the shaded veranda of their family home, the clients can see for miles over the Montana landscape.

their ambitions for an environmentally friendly compound as an act of reclamation. This approach has both limited disruption to the site and reduced the material resources needed for construction, while retaining the rural character of the ranch in a home that supports the traditional way of life implicit in the Montana landscape.

The ranch consists of the 'granary' and 'barn' clusters, located 1.6 kilometres (1 mile) apart on an unimproved road. The granary cluster makes use of a farmhouse (previously used as a stable) and granary that had been abandoned for more than fifty years. The architects kept the carcass of the farmhouse but completely reconfigured the interiors to create family bedrooms, a kitchen, a living area and a loft, which is used as a children's playroom. A central load-bearing wall, painted red, provides the spine and stairs of the building, and encases a fireplace and storage areas. Partition walls have been assembled from a mix of new timber and recycled wood from collapsed adjacent structures. Rather than hide these modern additions to create a historical replica, the architects stained the new timber and left the old in its original state, simply sealing it, so that the building tells its complex history through its physical appearance. The windows are double-glazed to provide protection against the cold, windy climate. Insulation has been added to the walls, and underfloor heating fitted beneath the reclaimed pine floors to create a comfortable and contemporary home.

The granary was in a reasonable condition, but one side of the building had been affected by the particularly vicious north wind that blows across the site. The architects therefore decided to lift the entire building and rotate it by 90 degrees; they then added new stained wood to the exterior and clear-sealed the old timber to protect it from the wind. Inside, they dismantled the six bins that had been used to store grain and created a laundry room, two guest rooms and a basketball court with a maple floor. Again, new and old materials are expressed with honesty, and the space has the modern comforts of underfloor heating, insulation and double-glazed windows. The water for both the granary and the farmhouse is drawn from an on-site well. To complete this cluster an L-shaped building containing a carport and workshop was assembled on site with a basic timber frame made from locally sourced materials.

Opposite: A decked veranda has been sensitively added to the guest quarters of the barn cluster, where the original logs still support the main walls.

Above: The outdoor privy has been replaced with a composting toilet and the original toilet seats transformed into a light fitting.

Right: One of the guest bathrooms demonstrates how modern fixtures and fittings have been successfully integrated within the original structures.

The barn cluster is the working part of the compound and comprises several buildings. A small two-cell granary has been adapted to provide guest accommodation equipped with a kitchenette and bathroom. A hundred-year-old log cabin has been sensitively converted into an office. The architects decided to pull down the dilapidated tin barn that held feed for the ranch's twenty horses through the long Montana winter. In its place they have constructed a polite but not historicized barn using traditional timber techniques with additional steel bracing to fortify the building against the north wind. Lastly, an outdoor privy has been replaced with a composting toilet, the architects reconfiguring the existing hand-carved toilet bench into a light fixture that now hangs above the new facilities.

Although it is the smallest building in the entire compound, the outhouse demonstrates the architects' dedication to the key philosophies driving this project: attention to detail; the combination of modernization with respect for the character and history of the buildings; and the minimization of ecological damage. The project itself demonstrates how existing rural buildings can be renovated and adapted to create a comfortable and efficient working ranch compound. The integrity of the buildings has been retained, while modern additions provide the next chapter in their history. The locals have shown real gratitude towards the clients and the architects for saving the buildings, which remain much-loved local landmarks. The project has thus been successful in integrating the clients with the community: neighbours call in to keep the family updated on local matters and to hunt on the grounds. By creating a modest and restrained home, the clients have gained the respect of the community and have been welcomed into it.

Opposite: Extensive use was made of timber reclaimed from collapsed buildings nearby. While new timber was stained, original timber was treated with a clear sealant.

Above and right, top: Double-glazing, insulation and underfloor heating were introduced to bring modern comforts to the old buildings, while traditional and natural materials are used in a contemporary and minimalist way.

Right: A general recreation space, with basketball nets allowing the owners and guests to shoot some hoops, was incorporated into the granary.

SITE AREA
7000 hectares (17,000 acres)

FLOOR AREA
House: 112 square metres (1200 square feet)
Granary: 112 square metres (1200 square feet)
Car barn: 116 square metres (1250 square feet)

ARCHITECT
Fernau & Hartman Architects
2512 Ninth Street, Number 2
Berkeley, California 94710, USA
+1 510 848 4480
general@fernauhartman.com; fernauhartman.com

PHOTOGRAPHY
Pages 94, 95t, 97t, 98, 101bl: Richard Fernau/Fernau & Hartman Architects
Page 99bl: Tim Hursley
Pages 95b, 96b, 99tl, r, 100, 101tl, tr: J.K. Lawrence
Pages 97b, 101br: Jim Wilson/New York Times/Eyevine

ROOF EXTENSION 'SYMBIONT FRIEDRICH'

MERZIG
GERMANY

The three young partners of German practice FloSundK are interested both in progressive forms of architecture and in responsible methods of urban planning. This project, a rooftop extension in the German town of Merzig, provided them with the opportunity to address these concerns. The challenge was to expand the top-floor apartment of a 1960s building in the heart of a built-up area. In response they planted two startlingly modern rectangular boxes on the roof, transforming the flat into a double-storey home and demonstrating how sites with unused potential can be developed into architecturally exciting structures to support increasingly dense urban populations.

The clients, a young couple, live at the top of the building, which belongs to their relatives and houses the family's long-established hairdressing business. Squashed into an 80-square-metre (860-square-foot) apartment, the couple desperately wanted more room for themselves and their plants. Money was tight, so they asked FloSundK to expand their living space and supplement their little north-facing

Right: The two lightweight timber boxes were assembled on the ground and lifted into place, requiring no reinforcement of the building's foundations.

Opposite: One of the boxes contains a winter garden; the other provides additional living space with sliding doors opening on to a roof terrace.

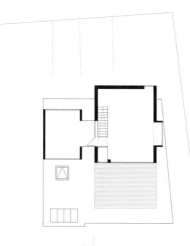 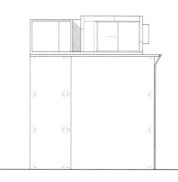

Left: Top-floor plan (left) and south elevation.

Below: The south-facing glazed doors of the rooftop living space bring in warmth from the sun during the winter and allow ventilation in the summer.

Opposite, left: The open-tread staircase allows natural light and cooling breezes to filter through the entire apartment.

Opposite, right: The living box juts dramatically out over the street, while the green concrete firewall of the winter garden adds a vibrant splash of colour.

balcony with a roof garden and enclosed winter garden on the smallest budget possible.

The architects began by reconfiguring the existing rooms, demolishing the wall between the kitchen and living area to create a large open space. In place of the wall they inserted an open-tread steel stairway, which punctuates the room and allows natural light to penetrate from the light-filled spaces on the roof. There they added two separate boxes – one for living and the other for the winter garden – generating 32.5 square metres (350 square feet) of additional living space, in addition to a 17.5-square-metre (188-square-foot) roof terrace. Constructed of lightweight timber panels, the boxes can be supported without any reinforcement of the existing

foundations, saving time, materials and money and causing as little disruption as possible to the building. The timber boxes were fully assembled on the ground and lifted into place with a crane.

Distinct materials reflect the different functions of the boxes. The living box, which extends dramatically beyond the edge of the existing building, is clad in dark-grey-painted chipboard, which has a high embodied energy but is made almost entirely from waste or recycled timber and is cheaper than plywood or solid timber. The winter-garden box is clad in zinc, which has the lowest embodied energy of all metals, is extremely durable and requires minimal maintenance. When left untreated, as here, the zinc takes on a natural grey–blue patina. On one face of the winter garden box, a green firewall of pre-cast concrete was screwed on to the existing concrete roof.

Both boxes are lined with 200 millimetres (8 inches) of insulation, and the glass windows and doors are double-glazed. In the winter the boxes benefit from passive solar gain through south-facing, fully glazed doors leading on to the roof terrace, with additional heating supplied by a wood-burning stove. In the summer the glazed doors open to catch cooling breezes that filter down to the lower floor through the open-tread staircase.

This original, economical and intelligent project has transformed a poky top-floor flat into a spacious, light-filled and energy-efficient home, preserving the original building intact while adding a new architectural chapter to its façade. The extension demonstrates how unused and often overlooked urban sites, such as flat roofs, can be built on without the need for demolition or the appropriation of virgin land. Such projects as this are essential to prevent urban sprawl and to enable high-density living. Extensions and houses that are fitted into the existing cityscape give people access to urban transport systems and such essential amenities as hospitals, schools and shops, supporting a less polluting lifestyle and preserving the green spaces that the earth so badly needs.

TOTAL FLOOR AREA
Extension: 50 square metres (540 square feet)
New total: 130 square metres (1400 square feet)

ARCHITECT
FloSundK
Bleichstrasse 24
66111 Saarbrücken
Germany
+49 (0)681 3799 710
info@flosundk.de
flosundk.de

PHOTOGRAPHY
G.G. Kirchner

HOUSE KOTILO

ESPOO
FINLAND

The architectural heritage of Finland is defined by timber buildings. From picturesque log cabins to magnificent eighteenth-century wooden churches, locally grown timber has long provided the backbone of the country's construction industry. The very first homes to be built in Finland were circular shelters made from thin logs, and this traditional aesthetic has been mirrored in the dramatic spiralling form of the House Kotilo (Gastropod House). Yet as well as drawing inspiration from the past, this wonderfully expressive timber building also addresses the pressing issue of climate change, providing possible solutions for the future.

Olavi Koponen was playing with a bar of polystyrene, bending and twisting it, when he had the idea of creating a house that spiralled around a central column. After winning a competition with the

design, he turned the model into reality by building the house for his own family.

Standing on a 300-square-metre (3230-square-foot) corner plot among a variety of residential buildings, House Kotilo is predominantly made from timber, with 237 square metres (2550 square feet) of living space. The walls, floor and roof are made from 180 prefabricated timber panels partly supported by a steel frame that allows the building to cantilever at the rear of the property, creating an imposing first-floor balcony and a shaded, decked area underneath. A fully glazed wall has been incorporated at the back of the balcony, while at the front of the building windows line a wedge that cuts into the curved ramp. It is an extraordinary and dynamic design.

The larch shingles that clad the exterior of the building, from a sustainable forest in Krasnoyarsk in

Siberia, have been left untreated and will age to a silvery grey. Unusually, a similar cladding appears inside, where a monolithic skin of Finnish aspen shingles spreads over walls and ceilings. A low-cost timber grown in sustainable plantations and forests in Finland and often used for interior joinery, aspen requires no treatment, and its muted colour and texture create a homely yet arresting interior.

On the ground floor, the open-plan kitchen and living areas encircle the central concrete core, which houses both the spiral staircase and an open fire. Bespoke timber furniture hugs the outside wall. The visitor burrows through these rooms, coming up for air only on reaching the first-floor bedroom, with its curious semi-open-plan bathroom at one end, and, at the other, the culmination of the spiral – the first-floor balcony. The floors are polished concrete speckled with pieces of blue recycled glass, and the staircase has been coated in shades of green rubber, a natural and renewable resource.

Koponen says that by using natural materials he wanted to re-create the simplicity, authenticity and cosiness of the vernacular buildings in which he spent time as a child. The building does indeed seem to wrap its occupants in a warm, tactile blanket, a feeling that is enhanced by its high levels of insulation: Koponen has lined the roof with 500 millimetres (20 inches) of rock wool and the walls with 200 millimetres (8 inches). The windows are triple-glazed with argon gas between the layers, and the house is fitted with a heat-recovery system

Opposite: With its spiral form and intricate larch cladding, this building provides a modern interpretation of the Finnish timber home, with walls, floor and roof constructed of wood panels.

Left, top: Wrapped around a central column, the spiral culminates at the rear of the house in a dramatic cantilevered first-floor balcony, supported by a steel frame.

Left, bottom: A concrete plinth protects the timber-clad exterior from damage through rainwater splashback, and, with the concrete core of the building, provides a thermally massive material to help regulate internal temperatures.

At the front of the house, a wedge-shaped void has been cut from the spiral and glazed to allow daylight to penetrate.

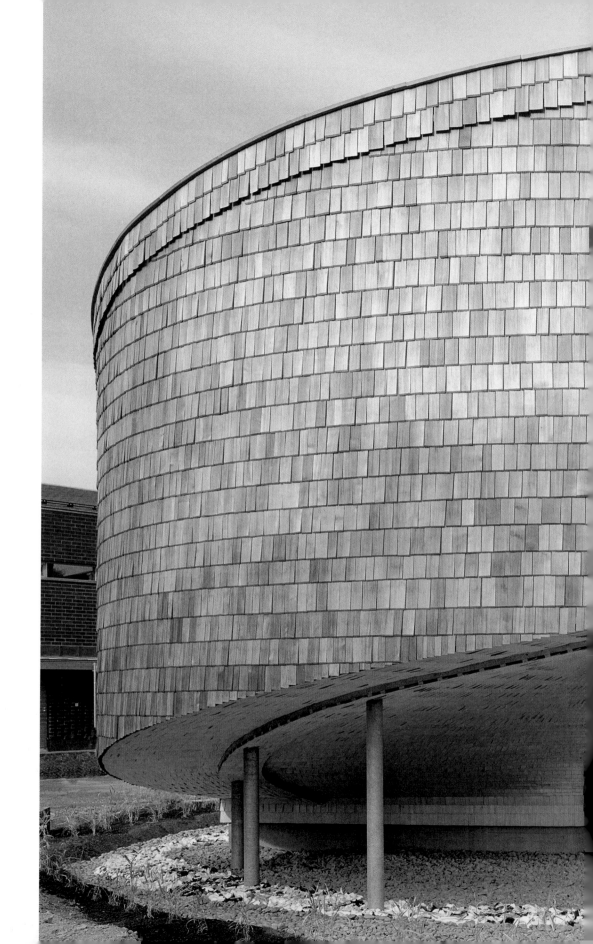

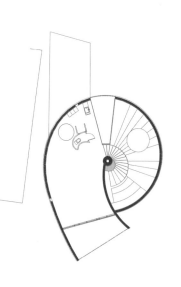

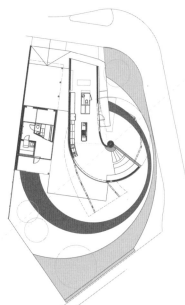

Above, from left: Section; first-floor plan; ground-floor plan.

Right: Insulation, triple-glazing and a heat-recovery system ensure comfortable and energy-efficient interiors. Much of the furniture has been custom-made to respond to the house's unusual shape.

Opposite, left: The internal walls and ceilings are lined with a tactile layer of local aspen shingles – a low-cost, low-maintenance, sustainable timber.

Opposite, right: The concrete core of the building incorporates an open fire and a spiral staircase, dramatized with recycled rubber in various shades.

that helps to cool the building in summer and heat it in winter. The small cuboid building next to the spiral houses a sauna – an almost obligatory feature in Finnish homes – heated by burning logs.

The house took thirteen months of careful and precise construction to complete, finishing in 2006. Many of the materials used – the timber, steel, rubber and glass – are recyclable, and the house's high levels of insulation ensure that it consumes minimal amounts of energy. Yet the most exciting aspect of the House Kotilo is its adventurous use of timber. Koponen demonstrates that although timber has been used for centuries, this extremely ecological material still offers endless possibilities when it comes to designing sustainable homes.

SITE AREA
300 square metres (3230 square feet)

TOTAL FLOOR AREA
237 square metres (2550 square feet)

ARCHITECT
Olavi Koponen Architects
Apollonkatu 23 B 39
00100 Helsinki
Finland
+358 (0)9 441 096
olavi.koponen@kolumbus.fi

PHOTOGRAPHY
Jussi Tiainen

OHANA GUEST HOUSE

NIULII
HAWAII
USA

The north shore of Hawaii's Big Island is famed for attracting the best surfers in the world. The Super Bowl of wave riding and the Vans Triple Crown of surfing are both held there during the winter months, when waves reach more than 9 metres (30 feet) tall. But waves mean wind, and the cliffs above these shores are constantly lashed by gusts of up to 130 kilometres (80 miles) per hour. It is a beautiful, rugged coastline, but not the most hospitable place to build a house. When Cutler Anderson Architects were asked to design the Ohana Guest House there, the wind dictated the form of the building. Yet it was not only the weather conditions that the architects took into consideration: they also crafted the building from locally sourced materials, rooting it both physically and aesthetically in the landscape.

The guest house occupies one of three lots on a 30-hectare (75-acre) compound near a former sugar mill in the community of Niulii. Despite the land's inclusion in a conservation area, it had been used for decades as a casual rubbish dump. When the clients, a retired executive and his wife, bought the plot, they became stewards of the land, spending six pains-taking years manually clearing away the junk (large equipment is banned by local ordinance). Only then could the project begin. The compound will eventually contain three buildings, with a large residence for the clients and the guest house providing accommo-dation for family and friends (*ohana* means 'extended family' in Hawaiian). In the interim the clients have made the guest house their permanent home.

To protect the occupants from the wind rolling off the Pacific Ocean, the house's U-shaped plan shelters

an outdoor courtyard that benefits from the southerly sun. A roof over the courtyard is tilted at a similar angle to the site and forces the wind up and over the property. The entire cedar-framed building is raised on a stone plinth, which derives from the *heiau*, a massive stone platform typically used for religious purposes in Hawaii. Made from a type of lava called A'a from nearby Parker Ranch – one of the largest and most renowned mountain ranges in the United States – the platform plays a crucial role in anchoring the lightweight frame to the ground. Steel ties laced through the timber rafters were pinned into the stone foundations, literally strapping the house to the land.

The sense of permanence conveyed by the stone platform contrasts dramatically with the transparent nature of the rest of the building. Extensive glazed panels were inserted in the cedar frame, naturally lighting the interiors and providing a 180-degree view of the coastline. Operable glass jalousies, screens and doors admit air to the interiors and allow breezes to circulate through the house. Coupled with the high thermal mass of the stone platform, this natural ventilation means the building can rely almost entirely on ambient cooling and heating. The only

Opposite: The house's slanted form is designed to deflect the ferocious winds that lash the coastal site.

Above and left: Glass is used extensively, bringing in natural light and revealing views, as well as opening to allow ventilation.

The lightweight cedar-framed building is literally tied to a sturdy platform of local lava, the heavy stone contrasting with the delicate glass superstructure.

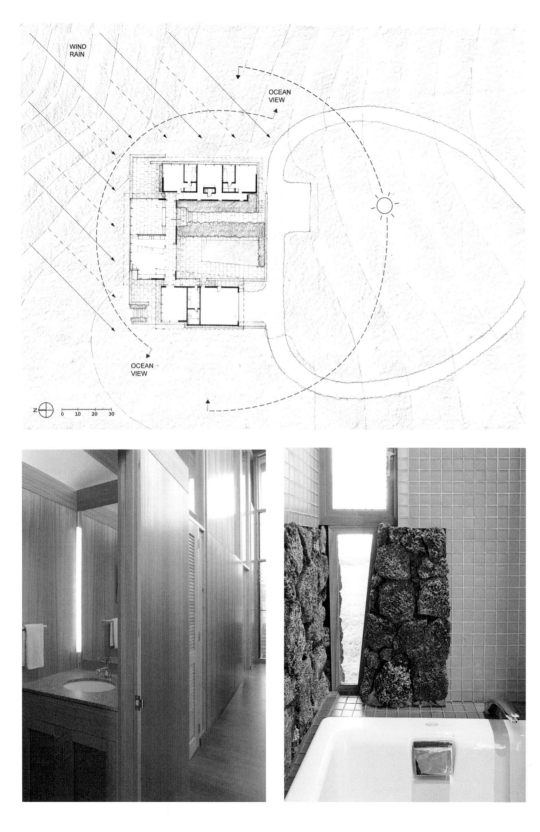

WIND
RAIN

OCEAN
VIEW

OCEAN
VIEW

N   0 10 20 30

exception is a small heating system that serves the library on the coldest of days.

The two wings of the 240-square-metre (2580-square-foot) house contain two bedrooms and a library on one side and a garage, utility and media room on the other. Linking them is the main living area. The clients were opposed to a large open-plan room, so the architects separated the living, kitchen and dining areas into discrete yet flowing spaces.

Inside, as outside, the building's natural materials provide the decoration. A low masonry wall bordering the rooms hints at the stone platform below, while the exposed timber frame and steel supports reveal the engineering that holds up the building. Stone paving and local sustainably sourced tamarind – a species of timber that was introduced to the island in the late eighteenth century – have been used on the floors. The furniture throughout the house, including the built-in kitchen, has been crafted from eucalyptus, another sustainable local timber. This is a house that has been pared down to its essentials and yet still looks impossibly chic.

With its wind-deflecting shape and rich palette of natural materials, this elegant house seems to belong in its setting. By using local stone and timber that were not processed and that travelled minimal distances, Cutler Anderson Architects have created a building with a low embodied energy. But, just as importantly, they have demonstrated how local and natural materials can be employed to create striking and novel architecture.

Above, left: Site plan.

Left: A simple palette of local materials, including sustainable tamarind and eucalyptus, was used to create the stylish interior.

Opposite, top left: The dining, kitchen and living areas share the same ceiling space but are discreetly divided using built-in furniture and timber partitions.

Opposite, top right: The stone platform, timber frame and steel supports are exposed in the interiors.

Opposite, bottom: The U-shaped building creates a private courtyard where the clients can enjoy the sun while sheltered from the wind.

TOTAL FLOOR AREA
240 square metres (2580 square feet)

ARCHITECT
Cutler Anderson Architects
135 Parfitt Way SW
Bainbridge Island
Washington 98110
USA
+1 206 842 4710
contact@cutler-anderson.com
cutler-anderson.com

PHOTOGRAPHY
Art Grice

MAISON JAVAUDIN

RENNES
FRANCE

Below and right: The front of the house was left untouched, but interior walls were removed, and a contemporary rear extension provides additional living spaces with extensive glazing carefully placed to draw natural light into the existing interiors.

Opposite: The lower floor of the extension is partly submerged in the ground, its concrete base and the surrounding earth bringing the benefits of thermal mass to the timber-framed addition.

From the driveway this house looks like any of the other stone buildings that line the streets of the French town of Rennes. But the back of the property offers something completely different. Architect David Juet has added a dramatic black extension to this family home, providing a new lease of life to the property and updating it to suit the living arrangements of his clients. By renovating an existing house to meet the demands of twenty-first-century living, he has also contributed to an important process by which the density of existing towns can be sensitively increased.

The extension was built to the French Haute Qualité Environnementale (HQE) standard, a mark of sustainable practice consisting of fourteen environmental criteria, including creating low-polluting buildings and following a responsible waste-management plan. Juet met the standard by incorporating low-energy materials into the design and by orientating the glazed areas to gain maximum passive heat and natural light from the sun.

Set on a concrete base submerged 75 centimetres (29½ inches) below the level of the garden, the extension's timber frame is generously insulated and clad in Trespa panels. Most of the large panes of double-glazed glass are operable, allowing cooling breezes to enter the house in the summer. In the winter, they admit light and warmth from the sun while the thermal mass provided by the concrete and earth that partly encase the ground floor also keeps the house warm. Additional heating is supplied by a wood-burning stove and a gas boiler.

The boxy form and extensive glazed areas are unashamedly modern and yet complement the

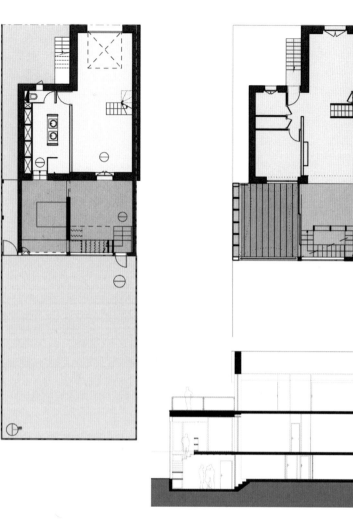

Above: Ground- (left) and first-floor plans; section.

Above, right, and opposite: Such details as the delicate stair balustrade and open-tread stairs allow natural light to penetrate.

existing building. Visually, the design is angular, like the silhouette of the roofline, and the modern materials offset the chunky old stones, while physically, the interiors have been knocked through to create larger, brighter spaces as well as to give easier access to the garden. Providing 115 square metres (1240 square feet) of additional space, the extension contains a ground-floor study area and master bedroom with en suite bathroom built into the existing basement. On the first floor the wall between the existing house and the addition was knocked through to create an extended living-room with sliding glass doors leading to a generous decked area, also accessible from the kitchen in the old part of the building. A third-floor deck caps the extension.

Drawing natural light into the heart of the building and opening many of the rooms on to the rear garden, the extension has also encouraged the family to spend a lot more time eating and relaxing outdoors. Responsibly designed and managed, this neat project has improved the energy-efficiency of the entire house, demonstrating how outdated buildings can be transformed into modern spaces that better capitalize on passive means for both heat and light.

FLOOR AREA (EXTENSION)
115 square metres (1240 square feet)

ARCHITECT
David Juet
36 rue du Maréchal Joffre
44000 Nantes
France
+33 (0)2 40 75 74 21
contact@kenenso.com
kenenso.com

PHOTOGRAPHY
David Juet

SUNSET CABIN

LAKE SIMCOE
ONTARIO
CANADA

Perched on the edge of Lake Simcoe in Ontario, Canada, Sunset Cabin is a building that both exploits and shows the greatest of respect for its setting. Taylor Smyth Architects designed the single-room sleeping cabin for a couple seeking solitude in the grounds of their family home. With four grown-up children, the clients found their house endlessly full of visitors, so they commissioned a retreat where they could escape the chaos and watch the sunset from their bed. Taylor Smyth's solution is a rectangular timber-clad box that complements rather than clashes with its surroundings. The building was designed to have as little impact as possible on the landscape during both construction and use, providing a space where the clients can relax in harmony with nature.

The 25-square-metre (270-square-foot) cabin consists of a glass box built within a timber box – the construction equivalent of the Russian doll. Three of

Opposite: A screen of cedar wraps the single-room cabin, regulating light and views, with large apertures overlooking the lake to the west, but few towards the main house.

Above, right: The house was constructed off-site, then disassembled and rebuilt beside the lake, cutting on-site construction time and waste of materials.

Right: Raised on caissons in a naturally level clearing, the cabin makes the lowest possible imprint on its setting.

the inner walls of the cabin are floor-to-ceiling glass, and an untreated-cedar screen encases the building. The slats on this screen have been carefully designed to provide views while protecting the privacy of the occupants: a large aperture in the west façade looks over the lake, and the density of the screen gradually increases towards the rear of the cabin, allowing the building to turn its back on the family house.

The cabin is entered from the south through an enclosed porch running the width of the building, providing a transitional space between the outside and the main room, which is the couple's bedroom. On the opposite, north, side of the cabin, the floor extends beyond glass doors to form a sheltered deck that provides access to an outdoor shower and composting toilet, which breaks down human waste into fertilizer. Both facilities are protected by the cedar screen.

The interior has been kept simple in order to maximize space, and is also made from exposed timber. The walls, floors and ceiling are constructed from birch-veneer plywood panels, as are the built-in bed and storage cupboards. The long glass wall that faces west to the lake sits at an angle to the cedar screen to allow the windows to be cleaned. Random gaps in the wooden shell create a wonderful play of

light on the glass, bringing natural and continuously changing decoration to an otherwise austere interior.

The site the clients chose for the cabin, 45 metres (150 feet) from the main house, not only has beautiful views but also is naturally level and devoid of trees, enabling them to preserve as much as possible of the existing landscape. The cabin is supported on two

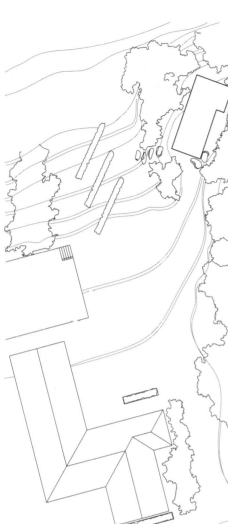

steel beams raised slightly by four concrete caissons, causing minimal damage to the surrounding vegetation. The choice of materials is also fitting to the site: cedar is a local timber with a tight grain and high natural oil content that make it extremely durable. It requires no maintenance and, left untreated as here, turns a silvery grey, helping the building merge with the woods behind. To the same end, the cabin's roof, which is visible from the main house on the hill behind, is planted with sedum and herbs, providing extra insulation as well as supporting biodiversity and replacing the vegetation now covered by the cabin.

The cabin was initially constructed in a car park in Toronto, allowing the details to be precisely calculated and the component parts prefabricated. The parts were numbered before being disassembled and transported by lorry the short distance to the site. The cabin was rebuilt in just ten days. Minimizing the construction time on site and the difficulties of working on a remote, sloping plot cut labour costs by 30 per cent. It also meant that less material was wasted on site and fewer materials had to be transported.

The clients now use the cabin all year round. The windows and doors are double-glazed with airtight cedar frames, and the roof, walls and floor are generously insulated. During the winter the building is heated with a wood-burning stove, with a small back-up electric heater that is called upon only during the coldest of evenings. In the summer the cedar screen provides solar shading, and breezes from the lake naturally cross-ventilate the space via the doors at the north and south ends of the building. The clients claim that the view provides them with endless entertainment, and they feel much more in tune with the seasons now that they have decamped to their private retreat.

Sunset Cabin is not a complicated structure. It uses basic materials to create an aesthetically dynamic and energy-efficient building. Little disruption was caused during construction because of the prefabrication, and with its muted cedar walls and plant-covered roof the cabin echoes its natural setting. Thoughtful additions, such as the wood-burning stove and composting toilet, mean the clients can enjoy the space while respecting the environment. The cabin treads lightly on its surroundings and, by doing so, preserves the beauty of this spot for generations to come.

Opposite, left: Site plan. The cabin (top) is situated a short distance from the family home (bottom).

Opposite, right, and right: With built-in bed and storage cupboards to maximize space, the interior is simply furnished, focusing attention on external views. The gaps in the cedar screen create an interesting play of light inside the cabin. A private deck outside the bedroom leads to a toilet and outdoor shower.

Below: The screen of untreated cedar – a local, durable and low-maintenance timber – integrates the cabin with the surrounding woodland.

TOTAL FLOOR AREA
25 square metres (270 square feet)

ARCHITECT
Taylor Smyth Architects
354 Davenport Road, Suite 3B
Toronto
Ontario M5R 1K6
Canada
+1 416 968 6688
info@taylorsmyth.com
taylorsmyth.com

PHOTOGRAPHY
Ben Rahn/A-Frame/Taylor Smyth Architects

RENOVATED HOUSE

CHAMOSON
SWITZERLAND

Architect Laurent Savioz has bravely transformed this rural Swiss farm building into a contemporary and sustainable home. He has not only restored a significant part of the original fabric, but also improved its thermal efficiency and equipped it with renewable energy sources, creating a building that is both appropriate to its context and environmentally sound.

The house was built in 1814 and sits on the edge of the town of Chamoson, surrounded by mountains. Its sturdy walls, constructed from the local limestone, have survived nearly intact, but it had been allowed to fall into a sorry state of neglect. Only part of the building was habitable, the roof leaked, and the timber cladding was battered beyond repair. Luckily, a couple living in a neighbouring property recognized the potential of the 380-square-metre (4090-square-foot) building and commissioned Savioz to transform it into a modern home. The brief was to create a simple, energy-efficient house filled with natural daylight while preserving as much of the building's original fabric as possible.

Savioz's solution was to insert a concrete frame within the stone carcass of the building. The concrete supports the existing walls and allowed the majority of the exterior envelope to remain intact. Largely hidden from view from the outside, the concrete is, however, left exposed where it replaces areas of original timber cladding (a feature of the local vernacular style). As with the rest of the building, Savioz here shows a commendable attention to detail, casting the exposed concrete in rough wood

Opposite: This sensitively renovated farmhouse – far more economical than a new build in terms of the resources used for construction – fits seamlessly into its mountainous surroundings.

Above: The original limestone walls of the house have been preserved by being supported with an internal concrete frame, which also provides insulation.

Left: Concrete cast in rough timber shutters evokes the texture and appearance of the original wooden cladding.

shutters to evoke both the appearance and the texture of the traditional timber. Vernacular architecture is given a modern twist.

The structural strength of the concrete gave Savioz the freedom to reconfigure the interior layout completely. One of the clients is an artist, and on the ground floor, as well as a sauna, wine cellar and utility room, there is a gallery where she can exhibit her work. A cast-concrete staircase leads to the first-floor kitchen, bathroom, living–dining-room and artist's studio, while the mezzanine level above provides space for a bedroom and en suite bathroom. Left exposed throughout, the concrete floors, walls and ceilings contrast both visually and texturally with the original stone partition walls encased within the building. Apertures of varying sizes were cut through these walls, allowing the spaces to flow into one another. The interiors feel monastic yet modern and, as with the exterior, Savioz has successfully combined the old with the new.

The project is inherently sustainable because it gives an existing building a new lease of life, reducing both the construction waste that would have been produced by demolition and the virgin materials that would have been required for a new build. Minimal materials were used throughout to great effect, and the preservation of the original walls has retained the character of the building.

But this house also has some surprising eco credentials. The outer walls and the roof are constructed from Misapor concrete, which includes an extremely effective insulation material. By adding a 300-millimetre (12-inch) layer of this concrete to the existing 600-millimetre-thick (24 inches) exterior walls, Savioz could dispense with any additional insulation material. Only in the roof did he need to add a 260-millimetre (10-inch) layer of glass wool to ensure the house's exceptional thermal efficiency.

Savioz retained the original openings that puncture the exterior walls and added some larger windows to let more natural light into the interiors. While all the windows have airtight aluminium frames and double-glazed panes, the larger openings are fitted with glass that reflects 80 per cent of solar

The original window apertures were retained and combined with new, large openings to draw light into the house.

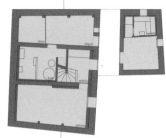
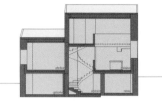

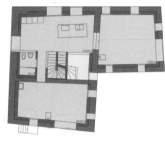

Above, left: Ground- (top) and first-floor plans.

Above, right: Section (top) and partial elevation.

Right: Exposed concrete walls dramatically carve up the interior.

Below right: An opaque glass screen divides the master bedroom from the en suite bathroom.

radiation, limiting the amount of heat absorbed by the concrete interiors during the summer months, when overexposure could cause uncomfortable temperatures. Ventilation is provided naturally through a heat-recovery system. On the roof 23 square metres (250 square feet) of solar panels power 35 per cent of the house's energy needs, and additional heating is provided by two wood-burning stoves, in the living-room and the artist's studio.

Together, these measures have ensured that the building meets the Swiss Minergie standard for sustainable construction, which requires that the building is efficiently insulated and airtight, that heat bridges are avoided and that an efficient airflow system is installed. As a result, the house's maximum energy consumption is approximately half that of a conventional new building. Savioz has not only transformed a neglected building into an architecturally daring house, but also created a home that has less impact on the planet than its predecessor.

Left: Generous openings cut from the stone and concrete walls allow the internal spaces to flow into one another. Here the sitting-room leads off the kitchen.

Below: The concrete frame allowed the architect to reorganize the interior, inserting a new mezzanine under the roof with a master bedroom and bathroom.

Bottom: The artist's studio is lit by an enormous picture window.

TOTAL FLOOR AREA
380 square metres (4090 square feet)

ARCHITECT
Laurent Savioz Architecte
Ch. St Hubert 2
CH-1950 Sion
Switzerland
+41 (0)27 322 5491
contact@loar.ch
loar.ch

PHOTOGRAPHY
Thomas Jantscher

PAYSAGE EN PALIERS

SERMÉRIEUX
ISÈRE
RHÔNE-ALPES
FRANCE

Architect Florian Golay was under considerable pressure when building this house in the Isère countryside to the east of Lyons in France. Not only were his parents the clients, but also the building was to be located on a tricky hillside plot. Golay's solution was to work with the natural gradient of the land to build a series of terraced interior and exterior living spaces that cascade down the hill. To achieve this, Golay combined heavy masonry concrete with a lightweight timber frame, exploiting the benefits of both materials to create a visually interesting, durable and impressively energy-efficient home.

The clients decided to buy the plot after becoming frustrated at the lack of interesting houses within their budget in the nearby town of Sermérieux. They asked their son to design a house that would take full advantage of the views and provide fluid indoor and outdoor spaces. The result is a contemporary building that faces south but spreads over the hillside to the west so that, from almost every room, the clients can step out on to a terrace or balcony.

On a stepped concrete plinth that cuts into the hillside to secure the house to the land, Golay has built a timber frame, two storeys high at the front and one storey at the back. The entire frame, including the roof, arrived on site prefabricated and was erected in three weeks. Golay chose timber partly for its ecological benefits but also because it is faster and cheaper to build with than concrete, and because in France reliable carpenters are easier to find than reliable bricklayers. The timber frame also allowed him to provide effective insulation – 150 millimetres (6 inches) of rock wool in the walls and 300 millimetres (12 inches) of cellulose insulation in the roof –

with a relatively thin wall, as insulation is packed into the frame rather than adding extra thickness, thus maximizing the livable floor area.

The exterior cladding of larch, a wood that is grown sustainably in this area of France, was left untreated, turning to a silvery grey. The only exception is the cuboid study, jutting out to the west midway between the lower and upper levels: it is clad in red render, giving the building a playful, contemporary feel. On its roof is a vegetable garden. The line of this red block is continued by a concrete terrace that is part garden and part swimming pool. A further, timber-clad deck is tucked behind the red block, leading directly from the first-floor kitchen. With a variety of private outdoor and semi-outdoor spaces, it is an arresting composition.

With a total floor surface of 150 square metres (1615 square feet), the interior is simple and open,

Opposite: Extensive use of timber, with a cladding of local, sustainably grown larch, integrates the house into its wooded hillside setting.

Left, top: The architect has dealt with the steep gradient of the plot by stepping the accommodation down the hill, with a cuboid study rendered in bold red jutting out between the lower and upper levels.

Left: The house's light timber frame is set on a concrete plinth, combining the ecological advantages of wood with the heat-storing capacity of concrete.

with two bedrooms and a bathroom on the ground floor, the study on an intermediate floor, and an open-plan kitchen, dining and living area on the first floor with a further bedroom and bathroom at the rear. During the summer all the windows can be opened fully to allow cross-breezes to ventilate the house. In the winter the double-glazed, low-emissivity windows, combined with the well-insulated walls and roof, prevent warm air from escaping. An automatic wood-pellet boiler provides the building's heating. The French government offers tax credits for half the cost of installing biomass boilers such as these, and grants are also available from local government for installing renewable forms of energy (these vary depending on the department, region or city).

The concrete plinth also plays an integral part in heating and cooling the building. Many airtight and lightweight timber-frame buildings have no way of storing heat energy and quickly become too hot or cold, relying on constant opening and shutting of windows or a mechanical heat-recovery system to regulate their internal temperature. The addition of a thermally massive material such as concrete overcomes this problem because it will slowly absorb and release heat energy.

Hybrid buildings that combine the ecological advantages of timber and the thermal-mass properties of concrete are becoming an increasingly popular way of creating thermally efficient buildings. In this case, the concrete is also an essential means of rooting the building to the hillside and creating a durable and comfortable house on a difficult site. It is a neat building and one with which Golay's parents, much to his relief, are extremely happy.

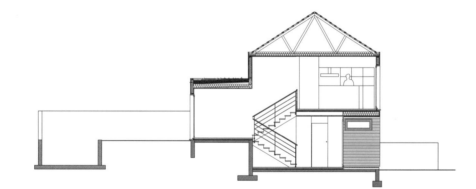

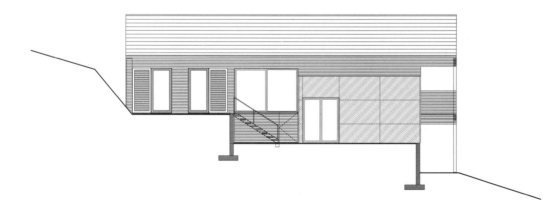

Left, top: Almost every room has access to a terrace or balcony.

Left: Section and west elevation.

Opposite, top: The open-plan kitchen, living and dining area benefits from access to both a south-facing balcony at the front of the property and a sheltered exterior deck to the west.

Opposite, bottom: The outside eating area is clad in larch, providing a sense of visual cohesion with the rest of the house's exterior.

TOTAL FLOOR AREA
150 square metres (1615 square feet)

ARCHITECT
Florian Golay and Christophe Séraudie
19 rue René Thomas
38000 Grenoble
France
+33 (0)4 76 84 91 55
golay.f@wanadoo.fr

PHOTOGRAPHY
Pages 132–133, 134t: Florian Golay
Page 135: Patrick Blanc

CASA EN LA FLORIDA

MADRID
SPAIN

This house by Madrid-based architects Selgascano takes the concept of site preservation to a new extreme. Dug into a plot on the outskirts of the city, it is designed to preserve in its entirety the vegetation that surrounds it. Not a single tree on the 2790-square-metre (30,000-square-foot) property was touched: instead, the house weaves dramatically to avoid them. Yet the building is in no way apologetic or polite. Rather than camouflage the house, the architects chose modern construction materials in vivid colours to distinguish it from its surroundings. The artificial and the natural coexist in harmony. This residence proves that efforts to conserve the natural landscape can lead to progressive and sustainable buildings.

The house was built for a couple and their two small children on a gently sloping plot in the residential district of La Florida. Before construction work began, the architects documented the precise location of every tree on the thickly wooded site, which included numerous species, such as oak, elm, ash, acacia and London plane trees, naturally germinated in random clusters from seeds scattered by birds and the wind. A kidney-bean-shaped void was identified among the trees, and it was there that the architects planted the house.

Semi-submerged, the building is divided into two wings, giving the architects greater flexibility when it came to bending the building to fit the clearing. A system of stepped terraces, partly decked in timber and partly covered by rock gardens, drops the visitor down to entrance level. One wing houses the living, eating and cooking areas and the other

Opposite: The house's irregular shape was designed to fit into an existing clearing on the heavily wooded site without disturbing any of the vegetation. Not a single tree was removed during construction.

Right: Semi-submerged, the house benefits from the temperature-regulating thermal mass of the earth, but is surrounded by clear acrylic above ground level, allowing in natural light.

Below: The division of the accommodation into separate living and sleeping wings joined by a corridor enabled the architects to adapt the shape of the house to its site.

The natural and the artificial coexist in stepped terraces of timber decking and rock gardens around the entrance. Some 200 extra plants were added to the site by the architects.

the bedrooms. The building's boundaries are irregular, but every angle and façade has been precisely calculated so that the house sprawls to within touching distance of the trees.

With all their projects, the architects take both an experimental and a sustainable approach to their choice of materials. The walls of this house are made from concrete, which bears the marks and texture of the pine shutters in which it was cast. The shutters were then reused to create part of the exterior decking. Only lightly sanded, they still show the patterns of the absorbed concrete, creating a beautiful contrast with the ipe wood also used for the decked surfaces. Above ground level a continuous clear ribbon of acrylic wraps around the building, providing durable windows and doors. The corridor that connects the two wings is made entirely out of this material, providing views right through the building to the decks on either side.

A dominating feature of the exterior is the roof, which is coated in brightly coloured rubber made from recycled tyres, orange on the public wing and deep blue over the sleeping quarters. Emerging from the orange roof are futuristic bubble-shaped skylights that illuminate the living area below, while large circular mats provide a place to sit, sunbathe or lie at night to stare up at the stars.

The 172-square-metre (1850-square-foot) interior is just as nonconformist as the exterior. The colour scheme continues: exposed steel ceiling beams are painted either orange or blue to create a sharp contrast with the crisp white walls; the bedroom and bathroom floors are lined with recycled orange rubber; and the living-room floor is covered in blue linoleum. The architects designed the angular fireplace, faceted orange steel bookshelves, a pendent fixture made from galvanized-steel tubing, and a pine dining table. Nearly all the remaining furniture was found in twentieth-century furniture shops, with pieces by Pierre Paulin and Charles and Ray Eames among the collection.

The architects' remarkable respect for the landscape entrusted to them has shown that, with imagination, a house can be manipulated to slot into an awkward site and still provide a comfortable and contemporary home. As well as preserving every single tree – the only casualty being an oak branch snapped by a concrete mixer – the architects have added a further 200 plants, ranging from ivy to a giant chestnut tree. As these grow, the house will become even more secluded, providing a hidden sanctuary among the trees. Equally importantly, a rare green lung has been preserved in this built-up area, absorbing carbon dioxide and releasing oxygen to improve the environment for the entire community.

Above, left: Site plan. The architects plotted the location of every single tree before beginning to design the house.

Above: The two wings of the building are defined by the colour of their roofs, made from recycled tyres: orange over the public living area and deep blue over the sleeping wing. Giant circular mats encourage the family to use the roof as an external living space.

Left: The main living area is filled with an eclectic mix of furniture, some made specifically for the house and other pieces found in twentieth-century furniture shops.

Above and below: Built-in kitchen units were specially made to fit the unorthodox space, and the architects designed an angular fireplace for the living area.

SITE AREA
2790 square metres (30,000 square feet)

TOTAL FLOOR AREA
172 square metres (1850 square feet)

ARCHITECT
Selgascano
Guecho 27
Madrid 28023
Spain
+34 (0)9 1307 6481
selgas1@selgascano.com
selgascano.com

PHOTOGRAPHY
Roland Halbe

GRYPHON EXTENSION

MELBOURNE
VICTORIA
AUSTRALIA

Every day people stop and stare at this house in the suburbs of Melbourne. Where once a modest single-storey, red-brick home stood there is now a jagged steel-clad structure with sharp angles, solar panels and oddly shaped geometric panes of glass. On closer inspection it can be seen that these dramatic lines form the shape of a gryphon – a legendary creature with the head, talons and wings of an eagle and the body of a lion. The architects Simon and Freda Thornton call this 'architectural fiction', a form of design comparable in literary terms more to the novel than to a textbook. The gryphon informed every design decision made on this project, from the shape of the roof to the positioning of the windows. But, like many good novels, this building also engages with current ethical and moral issues. Environmental sustainability provides its central storyline.

The clients were exacting when it came to the design of the house, and rejected six architects before they teamed up with Simon and Freda Thornton. One of the owners is a passionate environmentalist, and wanted the house to be a showpiece for sustainable architecture. The couple are also heavily involved in the organization of foster care and are foster parents themselves, welcoming an ever-changing group of children into their home. Part of the brief was therefore to create a house that would combine a sense of play and fantasy with that of nurture, support and homeliness.

The first decision was to save the front section of the existing property by upgrading its roof to make it more energy-efficient. The architects extended it both to the rear and to the side of the property. The frame and walls of the new building are made from Radiata pine grown in a sustainably managed plantation, with some steel parts to provide additional support. The walls are insulated with reflective foil and 75 millimetres (3 inches) of polyester bulk insulation, which is more than adequate for Melbourne's mild climate. The roof of the extension is constructed from a combination of zinc and aluminium laid on a 200-millimetre-thick (8 inches) layer of insulation and covered with a layer of steel. All the windows are double-glazed, their frames custom-made from sustainable cedar.

Inside the house every effort was made to use either natural or recycled materials. Australian recycled timber was used for the stair treads, bench tops, bookshelves and doors. The kitchen cabinets are plywood, and black recycled plastic was used for the doors. The floor of the extension is also plywood, with varied finishes including wool carpets, linoleum and ceramic tiles, while existing Baltic pine floorboards in the original property were re-stained, and reclaimed boards laid in the new entry hall. Non-toxic and VOC-free Bio-paint was used on all ceilings and walls.

The form of the building was manipulated not only to resemble a gryphon, but also to encourage passive cooling and heating through thermal mass and careful positioning of windows. In the summer, when the sun is at high altitude, overhanging eaves block solar gain. Where most of the internal walls are made from plasterboard, the stairs and fireplace are constructed from brick and finished with a hard plaster. At night the stairway and 'gryphon's head' (the highest part of the building) create a stack effect, drawing the colder evening air in through windows to cool the thermally massive core of the building. These high windows – the eyes of the gryphon – are currently opened using an electric motor, but the owners intend to automate this function using the house's 'smart wiring' system, which can be connected to differential thermostats and energy-monitoring devices. This will allow the system to work extremely efficiently without any intervention from the clients. In the winter, generous windows capture radiant heat from the sun, supplemented by warm air drawn to the kitchen via a duct from a spa enclosure next to the first-floor master bedroom. Hot water and heating are provided by a natural-gas-fired system connected to solar thermal panels

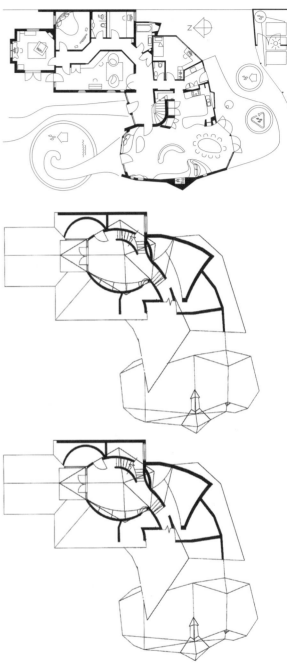

Opposite and left: The form of the gryphon is clearly distinguishable, particularly from above. The eagle's head sprawls across the original building, with wings of solar panels over the radical extension and a lion's tail curling into the sky.

Above, from top: Site plan; ground- and first-floor plans.

During summer nights the gryphon's head creates a stack effect, drawing air in through the windows (the gryphon's eyes) to cool the thermally massive core. The core then helps to regulate internal temperatures during the day, when the house is closed up to prevent the cool air from escaping. Beneath the pond at the front of the house is a huge rainwater tank.

Above, left and right: A spa enclosure is connected to the master bedroom by a shared exterior deck, with surplus heat directed in winter to the kitchen via a duct.

Left: The interior is as dynamic as the exterior, using natural and recycled materials and incorporating clever storage to maximize space, as in the staircase leading from the master bedroom to the peak of the gryphon's head.

on the roof. A further twenty-five photovoltaic panels are arranged to represent the feathers of the gryphon's wing and provide the house's electricity.

For use in the house, rainwater from the roof collects in a 22,000-litre (4840-gallon) tank submerged underneath the pond and timber decking in the front garden. Even the water run-off from the garden shed is collected in a separate 7000-litre (1540-gallon) tank. Waste water is reused in the garden and for flushing the toilets. The toilets and kitchen sink are connected to Melbourne's main reticulated sewage system, which the clients considered satisfactory, as it incorporates a sewage farm for sustainable waste processing.

The Gryphon Extension is equipped with many practical measures to reduce its impact on the

environment, but both the clients and the architects feel that the building's symbolism is equally important. The gryphon is a mythical figure that has always protected treasure. In the stories of old this treasure was precious metals and jewels; here, in a country where the changing climate is causing prolonged droughts, the treasure is the enormous water tank at the front of the property. The gryphon also represents a reuniting of the sky (the elements of air and fire, or the eagle) with the earth (water and earth, or the lion), after a long period of unbalancing in which the privilege accorded to the sky has literally disconnected the two in terms of the carbon cycle. Not only does this eye-catching house facilitate a greener way of life, but also it attracts people's attention and communicates to them the importance of sustainable design.

The curvaceous kitchen units were crafted from plywood and fitted with doors made from black recycled plastic.

SITE AREA
660 square metres (7104 square feet)

TOTAL FLOOR AREA
305 square metres (3283 square feet)

ARCHITECT
Simon and Freda Thornton Architects
24 Hanslope Avenue
Alphington
Victoria 3078
Australia
+61 (0)3 9486 3197
simonthornton@smartchat.net.au
thorntonarchitecture.com

PHOTOGRAPHY
Pages 142, 144–47: John Gollings
Page 143: Andrew Griffiths/lensaloft.com

LARCH HOUSE

MOSCOW
RUSSIA

Set up in 2002 by James McAdam and Tanya Kalinina, former directors of Alsop Architects' Moscow office, McAdam Architects has capitalized on the proliferation of new construction projects in Russia. The practice is involved in a number of major schemes in Moscow, including the high-profile development of the Red October Chocolate Factory. The Larch House is a smaller residential project in an affluent suburb to the north-west of the city, and is one of many houses built on newly privatized land that was previously part of a collective farm. The area has proved a magnet for the nouveau riche, resulting in the sporadic development of large houses in an array of architectural styles that pay scant attention to local culture and building methods or to sustainability. McAdam Architects have countered this approach by drawing on Russia's traditional styles and building techniques to create a house that, while undoubtedly modern and luxurious, also responds directly to Moscow's climatic, cultural and social environment.

Designed in a spiral form around a south-facing courtyard, the house mimics the layout of traditional Russian farmhouses, which were built in a U-shape around a yard or *dvor*, enclosed on its southern boundary by a low wooden fence. The Larch House also borrows the softwood timber frame typical of the traditional farmhouse, combining it with such modern innovations as a solid base of local brick and block and a 150-millimetre (6-inch) layer of mineral wool insulation. The windows are double-glazed with hybrid frames of timber, which provides a soft, domestic appearance, on the inside, and robust aluminium, which can withstand frost and snow, on the outside.

Opposite: The house gives a contemporary twist to the traditional U-shaped form of the Russian farmhouse, opening around a south-facing courtyard and turning its back to the cold northerly winds.

Above: Glazed south-facing walls open on to a sheltered courtyard, providing natural ventilation in the summer, while allowing solar heat and natural light to enter the house in the winter.

Above, right: The cladding is of stained Siberian larch, a plentiful local resource that is traditional to Russian houses.

Right: Minimal windows on the north-facing façade reduce the impact of cold winds. A bright-red entrance porch provides a focal point.

The house is clad in simple horizontal boards of Siberian larch, which grows in abundance in Russia and has been used for residential buildings there for centuries. It is a dense and durable wood with a high natural oil content that prevents it from warping or shrinking, making it ideal for exterior cladding.

On the north, east and west façades of the house the larch cladding, stained with a grey, semi-transparent, water-based paint, is almost entirely solid, with minimal window openings – another feature adopted from traditional local buildings, which effectively turn their back on the harsh Russian wind. By contrast, the protected southern façade is generously

Right: Ground- (left) and first-floor plans.

Below: The house is simple and minimal but also luxurious, with several expansive bathrooms.

Below, right: The main entrance leads into an impressive double-height atrium, with a central staircase leading to the first floor.

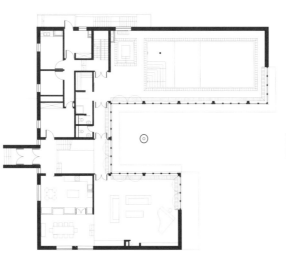

Above: The first-floor roof terrace over the swimming pool provides a private outdoor space.

Above, right: The generous living spaces are enhanced with rich natural materials such as dark slate for the fireplaces and solid oak for the floors.

glazed, allowing low sunlight to enter the house during the winter, while protruding timber canopies cut out direct sunlight in the summer. In warmer weather, glass doors around the courtyard can be opened to extend the living areas into this central space and allow cool breezes to ventilate the home.

Entered via a red-glazed brick arch at the centre of the flat street façade, the house is minimalist and airy inside with solid oak floors, white plastered walls and dark slate fireplaces. A large living area, dining-room, kitchen and service areas are on the ground floor, with four bedrooms and a study above. The main spaces are arranged in the taller wing to the east; the lower, western wing houses a swimming pool with a first-floor terrace above.

From the outside, this is an unpretentious building with a simple form and humble skin. McAdam Architects have learned lessons from the past and sourced materials locally, including Siberian larch, a natural and renewable resource. The architects have created a house that can protect itself from the harsh extremes of the Russian climate by exploiting radiant heat from the sun on its southern façade while minimizing the impact of the winds from the north. But this is also a luxurious home. It provides 1200 square metres (12,900 square feet) of living space and has all the trappings of a wealthy household, including a pool and service areas. McAdam Architects have demonstrated how Russia's architectural heritage can be adapted to provide contemporary environmentally conscious and comfortable homes. Such projects as this are essential if Russia's traditions are to be preserved in the construction frenzy that is currently transforming the country.

SITE AREA
18 hectares (45 acres)

TOTAL FLOOR AREA
1200 square metres (12,900 square feet)

ARCHITECT
McAdam Architects
Unit 11, Union Wharf
23 Wenlock Road
London N1 7SB
UK
+44 (0)20 7253 1738
info@mcadamarchitects.co.uk
mcadamarchitects.co.uk

PHOTOGRAPHY
Pages 150br, 151l: Ivan Boiko
Pages 148–49, 150bl, 151r: Project Russia

STONE HOUSE

SHIMANE PREFECTURE
JAPAN

The roof of the Stone House rises triumphantly from a pile of rubble, like the survivor of a landslide coming up for air. It is a bizarre sight: a house built and then partly buried. But by abandoning architectural convention, Sambuichi Architects have created a home that exploits the properties of its natural materials to the full. These materials, not only natural but also recyclable, are used as sparingly as possible to create a shell that can passively regulate internal temperatures. Embedded Tardis-like in the stone, this clever house is beautifully crafted, with bright and eclectic interiors.

The clients are a family of four, who asked the architects to create a house that could be used throughout the year, and a guest annex with a large terrace. Located on the border between Hiroshima, Yamaguchi and Shimane prefectures, the house sits on a plateau surrounded by rice fields in a mountainous region with heavy snowfall, famed throughout Japan for its ski resorts. In the winter the house is fully exposed to harsh northerly winds and in the summer it is engulfed by the hot, humid air that settles over the mountains. The challenge Sambuichi Architects set themselves was to create a building that could cope with these harsh climatic conditions without relying on constant heating or air conditioning.

They decided to bury the timber-framed building in crushed stones, reducing the surface area exposed to the outside air and using the thermal mass of the stones to control internal temperatures. In the snowy winter, air trapped between the stones creates a natural layer of insulation. In the summer, air is cooled as it passes through the crushed stones

Above: The elegantly executed timber-and-glass structure is embedded in a pile of stones, which provide thermal mass to moderate naturally the area's extreme temperature fluctuations.

Above, right: Cutaways in the rubble, created with the help of metal caging, create an entrance and allow natural light to reach the ground-floor interiors.

Right: The slanted glass roof prevents snow from settling on the house during the winter, while a layer of timber slats on the underside of the glass controls the amount of natural light entering from above.

Internally, rooms are arranged to make the most of passive energy control. Heavily used areas are located on the stone-covered ground floor, which is warmer during the winter months and cooler in the summer.

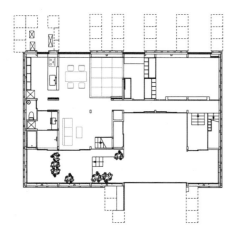

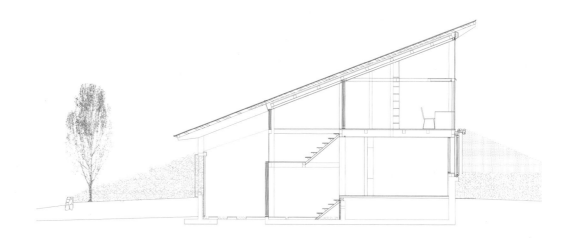

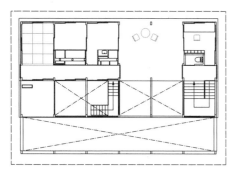

Above: Ground- (top) and first-floor plans.

Above, right: Section.

Right: The multifunctional first-floor terrace is warmed through its glazed façade in the winter and can be cross-ventilated in the summer, helping to cool the entire building.

before reaching the interior of the building. It is a simple and effective solution.

To minimize the roof area, the architects combined the main house with the guest annex in a single 271-square-metre (2920-square-foot) space. The interior was then configured according to the way each space would be used and how that would affect the temperature, light patterns and natural flow of air through the building. The guest areas, which are occupied much less frequently than the main family spaces, are located on the first floor, along with a study and small and large indoor terraces. The rooms at this level sit above the line of rubble and benefit from glass curtain walling across the entire south-

facing elevation. The large terrace is multifunctional: a solarium in winter, clothes-drying space during the rainy season and entertainment space when required. It traps warm air in the winter and can be cross-ventilated during the summer by opening doors on the east and west façades. This helps maintain a comfortable temperature throughout the year in the lower levels of the house, as the lack of partition walls and the open staircase mean that air can circulate freely. The ground floor is divided into a large communal sleeping room, family room, living and dining area, kitchen, bathroom and generous hall. Two more private terrace areas are enclosed by the stone walls on the north face of the building. All the interiors are simply clad in timber, with limestone, traditional *tatami* mats and wood floors.

For the roof, the architects chose glass because snow does not settle well on it, especially when it is used as it is here, slanting upwards to create the glazed expanse of the southern elevation. A layer of irregular timber slats was attached to the underside of this glass, each slat carefully positioned and angled to control the natural light entering the rooms. The

On the northern side of the house a covered entrance courtyard and raised seating area have been tucked between the mass of stones and the timber walls of the building.

slats also control the movement of hot air rising through the building, creating an effective layer of insulation without any additional materials.

It is the simplicity of this project that makes it so exciting. The architects have used elementary materials – stone, timber and glass – in the most minimalist way possible to create a building that can withstand the harshest of weather conditions. It is economical in both its use of materials and its dependency on fossil fuels, and will be fully recyclable at the end of its life. Visually, the combination of the rubble base with the soaring roof and delicate glass walls is dramatic and yet contextually fitting. The architects have demonstrated a true understanding of the properties of their materials to create an eloquent example of a site-sensitive and environmentally aware home.

SITE AREA
700 square metres (7535 square feet)

TOTAL FLOOR AREA
271 square metres (2920 square feet)

ARCHITECT
Sambuichi Architects
8-3-302 Nakajima-Cho
Naka-Ku
Hiroshima 730-0811
Japan
+81 (0)82 544 1417
samb@d2.dion.ne.jp

PHOTOGRAPHY
Hiroyuki Hirai

FOCUS HOUSE

LONDON
UK

Focus House is an unashamedly modern building that uses progressive materials to create an energy-efficient family home. Frustrated with living in a high-maintenance and inflexible Victorian end-of-terrace house, the client commissioned Bere Architects to create an economical and easy-to-maintain home on the adjacent triangular plot. The resulting property is a daring metallic building that uses its site to the maximum but minimizes the occupants' demands on the earth's resources.

The house is constructed from stacked metal-clad blocks punctuated with sizeable glazed areas. This cascading box arrangement maximizes light while neatly demarcating the internal spaces. The west elevation, which looks out on to the street, is a mere 2.8 metres (9 feet) wide and politely slots in next to its Victorian neighbour, but at the rear the building spreads out to the boundaries of the site, reaching a width of 7 metres (23 feet). On the ground floor an open-plan living, kitchen and dining area opens with sliding doors on to the garden. A short flight of stairs leads to the first floor, comprising two children's bedrooms, a bathroom and a study, which juts out dramatically over the front entrance. The second floor is reserved for the parents and contains a master bedroom and bathroom with commanding views. Lighting fixtures and extensive storage units in the internal walls, coupled with generous ceiling heights and bright finishes, make the 250-square-metre (2690-square-foot) house feel larger than it is.

The basic building material is cross-laminated timber slabs that look like giant lengths of plywood. Strips of Austrian spruce are glued crosswise on top

of one another to a thickness of 200 millimetres (8 inches) using a solvent- and formaldehyde-free adhesive, producing incredibly strong panels that can span larger distances than conventional timber. Used for wall, floor and roof slabs, the panels were prefabricated in Austria with window and door openings factory cut, enabling the building to be constructed and fully fitted out in just six months.

The architects decided to build with wood because of its low embodied energy and because timber continues to act as a carbon store after felling. They have estimated that during its lifetime the timber frame will have removed 42.37 tonnes of carbon dioxide from the atmosphere, where an ordinary Portland cement structure would add 32.42 tonnes of carbon dioxide to the atmosphere through the burning of fossil fuels during its manufacture. Transporting the wood by lorry from Austria caused 2.97 tonnes of carbon dioxide to be emitted, but the statistics are still impressive.

The remaining construction materials were also selected for their environmental performance. The concrete of the slab and foundations contains 70 per cent GGBS (ground, granulated blast-furnace slag), a by-product of iron production. The frame is lined with 200-millimetre-thick (8 inches) sheets of Foamglas insulation, and the whole is clad in a skin of zinc, which has the lowest embodied energy of any metal and is 100 per cent recyclable, as well as being durable and requiring little maintenance.

The design of the Focus House draws on many PassivHaus principles. As well as being extensively insulated, the house is airtight thanks to meticulous detailing and high-quality Scandinavian windows that

Opposite, left: The walls, floor and roof of the house are all made from slabs of timber, which continues to act as a carbon sink long after it is felled. The cladding is of zinc, which is durable and recyclable.

Opposite, right, and below: Spreading from 2.8 metres (9 feet) at the front to 7 metres (23 feet) at the back, the house takes full advantage of its triangular site, using a hidden pocket of urban space and helping to prevent urban sprawl.

are double-glazed and timber-framed. A heat-recovery system channels fresh air into the building while helping to regulate the internal temperature and minimizing the need for additional heating and cooling. Solar thermal panels on the south elevation generate on average 50–60 per cent of the house's hot water requirements, varying from 100 per cent in the peak summer months to 5 per cent in the depths of winter.

Despite its radical appearance, Focus House was submitted for planning permission with many letters of support from neighbours, and met with no resistance from planning officers. It demonstrates how modern techniques and materials can be used to create visually innovative buildings that use minimal energy both in their construction and in their occupation. High-density urban centres are now being seen as crucial to the long-term survival of the planet, since they preserve precious green country-side and cut down on the distances people and goods have to travel. Such inventive projects as the Focus House, exploiting every last awkward scrap of redundant urban land to create spacious dwellings, will help make this a reality.

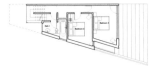

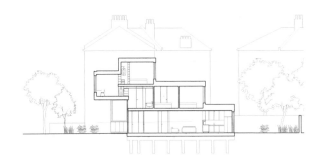

Top, left and right: The staggered arrangement of the house's box-like elements allows for more windows, bringing in natural light at every level, while the boxes delineate different areas within the house.

Above, left to right: Ground-floor plan; second- (top) and first-floor plans; south elevation (top) and section.

The ground-floor living and kitchen area is one continuous space entered at the front door and exited via large sliding doors into the garden. Good insulation, double-glazed windows, solar panels and a heat-recovery system all make the house easy and economical to maintain.

TOTAL FLOOR AREA
250 square metres (2690 square feet)

ARCHITECT
Bere Architects
24 Rosebery Avenue
London EC1R 4SX
UK
+44 (0)20 7837 9333
bere@bere.co.uk
bere.co.uk

PHOTOGRAPHY
Peter Cook/View

VILLA BIO

LLERS
SPAIN

Villa Bio marks a new chapter in organic architecture.
Designed by Enric Ruiz-Geli, founder of architectural
practice Cloud 9, the building demonstrates how
modern technology can create a home that connects
with the environment through both its form and the
way it functions. Set in the suburb of Llers, just north of
Barcelona, Villa Bio challenges every preconceived
notion of the polite suburban home. The classic boxy
Spanish villa has been replaced with an undulating
snake-like form, with plants sprouting from the roof
and large panes of glass lurching towards the street,
demolishing the division between public and private
space. It is out of step with its neighbours and yet, seen
in a broader context, gels with its natural surroundings.

The house takes its form from the topography of
the site, in particular the stepped, forested area
behind it. Its spiral shape digs into the ground to form
a submerged garage, then twists back towards the
rear of the site, before turning again and rising to an
impressive cantilevered section that projects towards
the street. This undulating form is made possible
by the use of reinforced concrete. Left exposed on
the long north and south façades of the building, the
concrete panels are ingrained with swirling patterns
generated by transforming a three-dimensional
image of the surrounding landscape into a mould, in
which the concrete panels were set. The resulting
walls are heavily textured and graphic, and add a
sense of movement to the exterior façades.

In contrast to the windowless elevations, which
shield the house from its neighbours, the east and west

façades are glazed with vast, uninterrupted panes, and this glazing continues on the south-facing wall of the upper ramp, allowing light in. The building has a roof garden consisting of a 70-millimetre ($2^3/_4$-inch) layer of volcanic earth planted with aromatic species. As well as extending the line of the sloping hillside site, the roof has many practical advantages: lessening water run-off, helping to insulate the house and protecting the building from the Tramontane, a strong wind that blows in this region. The roof also augments outdoor space on a small plot with little garden.

Various incursions were made into the fabric of the building to control the internal climate and light levels. The roof is scattered with solar tubes that cut through the concrete ceiling and introduce light into the interiors, while three large circular skylights are fitted with transparent filter glass to cut out 40 per cent of solar radiation. This natural light is supplemented by electric lights equipped with energy-saving sensors that reproduce the chromaticity of daylight. To allow cross-ventilation, holes were drilled through the concrete sides of the building and capped by glass stones, which both limit the amount of wind entering the house and create an interesting spectacle at night, when they are illuminated by escaping light.

The interior is as dramatic as the exterior. The client asked for a home without stairs to accommodate his two young children and disabled father. The ground-floor bedrooms and main living spaces can therefore

Opposite: The house echoes the undulating form of the hill to its rear with a planted roof that insulates the building as well as supporting biodiversity outside.

Above: At the rear of the property large, slanted panes of glass allow the family a constant view of the garden and the forest beyond, as well as drawing in natural light.

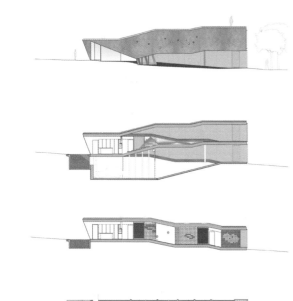

be reached by a gentle ramp. Sculptural timber steps provide access to the top-floor master bedroom, as well as an enclave for the dining table and a bench at ground-floor level facing the wood-burning stove in the living-room. Surfaces are bare, with exposed-concrete floors and walls throughout.

Villa Bio has attracted considerable attention from the family's neighbours. During construction it was called such derogatory names as 'the bunker' and 'the ski ramp'. Now it is complete there has been a gradual shift in attitude, and neighbours who have been invited up to the roof garden are amazed by the amount of greenery and the views towards the Pyrenees.

By combining his love of technology with his awareness of nature, Ruiz-Geli has transformed this tight site into a daring yet comfortable family home that is opening people's eyes to the benefits of contemporary sustainable design. It is no longer enough for a building simply to connect visually with its surroundings; by incorporating modern materials and energy-saving techniques, a building such as this can now help to sustain the wider environment.

Left, from top: South elevation; sections; ground–first-floor plan.

Below, left: Light is drawn in via a glazed south-facing wall in the upper ramp and solar tubes in the roof that cut through the concrete ceilings.

Below and opposite, bottom: The use of reinforced concrete for the shell enables a dramatic cantilever towards the street as well as bringing the advantages of thermal mass.

Opposite, top left: Natural ventilation via holes drilled in the side walls is moderated by glass caps.

Opposite, top right: Inside, a gentle ramp leads from the front door to the living area; timber steps lead up to the master bedroom above.

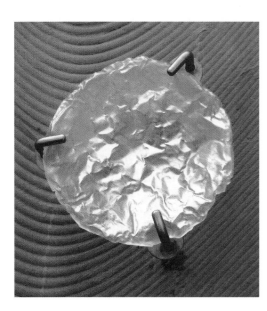

ARCHITECT
Cloud 9
Passatge Mercader 10 bajos 32
08008 Barcelona
Spain
+34 (0)93 215 0553
info@e-cloud9.com
e-cloud9.com

PHOTOGRAPHY
Pages 164, 165tr, b: Gunnar Knechtel
Page 163: Aniol Rescloso
Pages 162, 165tl: Enric Ruiz-Geli

SPLIT HOUSE

YANQING
BEIJING
CHINA

Wedged into a valley that runs alongside the Great Wall of China, the Split House is part of the Commune by the Great Wall – a novel hotel complex comprising thirty-two villas and ten contemporary chalets, designed by a number of well-known Asian architects, including Japan's Shigeru Ban and Thailand's Kanika R'kul. However, it is much more than just a hotel villa. The architects, Atelier Feichang Jianzhu, used this commission as an opportunity to develop a new form of environmentally aware architecture that could be more broadly applied in China. They created a house that is not only made from sustainable materials, but also flexible in form, allowing it to be built on various rural sites. In the context of China's ongoing economic boom and its vast and rapid explosion in building, the relevance of the Split House becomes compelling.

As its name suggests, the house is a single volume split in two. Its form is inspired by the traditional Beijing courtyard house, the outdoor courtyard here being closed off by the surrounding mountains. This blurring of the boundary between nature and architecture is continued by a natural stream that was discovered on the site and re-routed to meander through the courtyard and flow underneath the glazed entrance lobby, which fuses the two separate wings.

By splitting the house in two, the architects have separated the public living areas from the private bedrooms. But this division also gives the building its flexibility: the angle between the two wings could vary according to the site and the preferred living arrangements of the occupants. Here, the two wings were set at an acute angle determined by the existing trees

Opposite: The house provides a flexible prototype of sustainable architecture, with two wings that can be arranged in a variety of configurations according to their setting.

Right: Here the two arms are arranged in a V, the angle carefully calculated to preserve trees in the central courtyard, through which a natural stream flows.

Below: A glazed entrance lobby connects the two separate wings of the building, one housing public living areas and the other more private bedrooms.

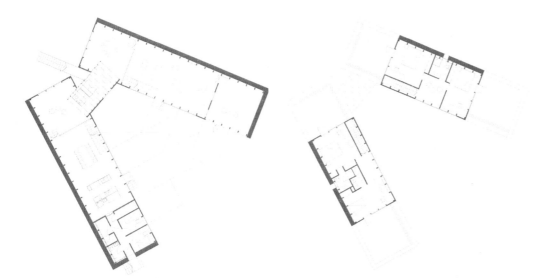

Left: First- (left) and second-floor plans.

Below: The house's traditional and sustainable materials – timber and rammed earth – are left exposed inside. As well as providing good insulation, rammed earth is made from clay soil, which can usually be locally excavated.

on the site. But the wings can be placed side by side, in parallel, at right angles, back-to-back: the possible configurations are manifold. The split also provides the opportunity to use only half of the house when the number of residents is low, reducing energy demand.

The wings fuse modern and traditional materials. Expansive glass curtain walling was installed on both the ground and first floors, but the main structure is a laminated plywood frame with rammed-earth walls; timber and rammed earth are the primary traditional construction materials in China. Made by pounding and compacting soil with a high clay content into hard blocks, rammed earth is a natural material with a low embodied energy. In the case of the Split House, the soil was excavated from the project site and other nearby construction sites; the same could be done for countless projects throughout China. Rammed earth also has an extremely high insulation value and, if the building were ever abandoned, the walls could be left to disintegrate naturally, generating no waste material and causing minimal permanent damage to the landscape. By selecting this historically resonant material for such a high-profile and contemporary project, the architects have not only publicized its sustainable credentials but also reintroduced it as an acceptable material for modern Chinese buildings.

Inside, the building's fabric was left exposed, with the wooden frame and earth walls juxtaposed against timber and stone floors. Yet despite this natural palette of materials, the 450-square-metre (4845-square-foot) interior reflects a European influence: living areas are open-plan with generous floor-to-ceiling glazed areas and a scattering of designer furniture. The house may be made out of mud, but it has a sleek, contemporary aesthetic.

The sudden rush to build in China has resulted, in many cases, in a reckless embrace of all things Western. This has led to buildings that are aesthetically at odds with their surroundings, using such materials as concrete to produce fast and economical results. The definition of contemporary Chinese architecture and its approach to sustainability is therefore yet to be resolved. But with such architects as Atelier Feichang Jianzhu producing work that refers to the latest international trends while also drawing on China's rich building heritage, a contemporary vernacular style is beginning to emerge. The Split House proves that sustainability is very much part of this agenda.

Traditional building materials are combined with extensive glazing, modern furniture and a contemporary form to create a dynamic building.

TOTAL FLOOR AREA
450 square metres (4845 square feet)

ARCHITECT
Atelier Feichang Jianzhu
Yuan Ming Yuan East Gate Nei
Yard No. 1 on Northside
Yuan Ming Yuan Dong Lu
Beijing 100084
China
+86 (0)10 8262 2712
fcjz@fcjz.com
fcjz.com

PHOTOGRAPHY
Pages 167–69: Shu He
Page 166: Fu Xing

HARRIS RESIDENCE

BELVEDERE
CALIFORNIA
USA

Light, bright and with sleek contemporary lines, the Harris Residence encapsulates chic Californian living. It has generous walls of glass, modern minimalist furniture, tactile surfaces that tempt trailing hands and the most astonishing view across the Belvedere lagoon to Mount Tamalpais. It is difficult to believe that when the Harris family bought the house in 2003 it was a dingy, dishevelled mess. Even less obvious is the fact that this transformation has been achieved to the highest sustainable standards. Every decision made in this renovation project, from the solar panels on the roof to the PVC-free blinds, has been driven by the clients' green agenda. The result demonstrates that it is possible to be sustainable without sacrificing style.

The original timber-framed house was built in 1949 and had been poorly remodelled by subsequent owners. The Harrises' aim was to create a home that reflected their strong values. Mr Harris is a veteran environmental and political activist who co-founded a luxury hybrid-car service that, among other things, takes celebrities to the Oscars; Mrs Harris is a fashion designer who wanted a contemporary home that connected freely with the outside. The couple asked San Francisco-based architect Christopher C. Deam to help them orchestrate this ambitious plan.

Over the course of two years, 80 per cent of the building was dramatically transformed. Interior walls were ripped out and glazed areas installed to create an open-plan sequence of rooms on three sides of a central courtyard. Private bedrooms for the couple and their two young sons were tucked around the

Pages 170–71: Sitting on the edge of the Belvedere lagoon
and enjoying views across the water to Mount Tamalpais,
as a renovation this project consumed fewer materials and
less energy in its construction than a new build.

Above: The original 1940s house has been completely
reconfigured around a central courtyard planted with
native drought-resistant plants, and is fuelled entirely
by solar panels on the roof.

Right: Site plan.

Opposite, top: The house turns its back to the neighbours,
with an enigmatic street façade.

Opposite, bottom left: Narrow strips of sustainably sourced
bamboo on the floors visually extend the interiors. All the
wood in the house is recycled or sustainable.

Opposite, bottom right: The bathroom typifies the house's
interconnection between interior and exterior and its
combination of the contemporary with the natural.

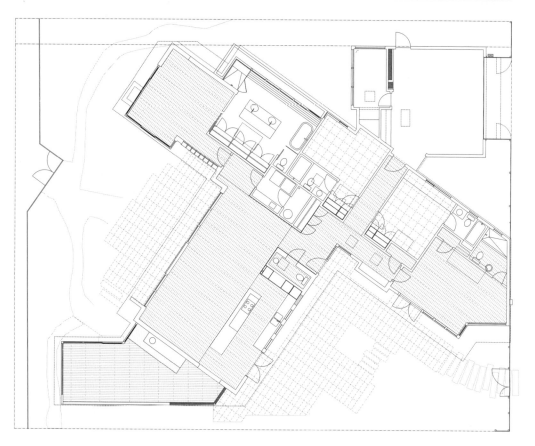

back. In place of the overspilling skips that usually accompany renovation projects, a responsible waste-management system was put in place: the old foundation and formwork was crushed into aggregate to be used in other building projects; old doors and windows were donated to a non-profit-making architectural salvage yard in San Francisco; and such original elements as plumbing fixtures and timber were recycled and reincorporated into the building.

Materials added to the house were selected for their sustainable credentials. If the required timber (including ipe, redwood, Douglas fir and eucalyptus) could not be salvaged, it was bought from sustainably managed forests and plantations, as in the case of the bamboo that covers most of the floors, its thin board width serving visually to elongate the central living space and corridors. Since concrete manufacture is heavy in carbon dioxide emissions, for floors that need to be hard-wearing, as in the garage, the concrete was mixed with 30 per cent fly ash. Low-VOC paint has been applied to the ceilings and many of the walls, including an exposed-brick wall in the living area, previously hidden behind drywall panels. The counter that demarcates the kitchen from the living area is wrapped in a monolithic skin of marble, a hardwearing and natural material.

The 430-square-metre (4630-square-foot) building is also extremely energy-efficient. UltraTouch cotton insulation was applied in a 140-millimetre (5^1/$_2$-inch) layer in the walls and a 250-millimetre (10-inch) layer in the roof. Insulating glazing has been used on the windows and doors, as well as low-emissivity coatings, while adjustable solar shades maximize solar gain in the winter and block out excessive radiant heat in the summer. An underfloor heating system has been installed, and the boiler and all other appliances in the house have high efficiency ratings. The minimal amount of energy required to run the house is generated by solar panels mounted on the roof. Since the family moved in they have channelled hundreds of excess kilowatts of power back into the grid.

The building code stipulated that all water run-off from the roof and landscape be diverted to the lagoon, so the clients were prohibited from installing a water-collection system, but they have designed the garden with native Californian drought-resistant plants, which sway in the breezes that come off the lake to ventilate the house naturally. The only section of the garden that needs irrigating is a raised bed

The house uses passive natural means to cut energy consumption: opening, low-emissivity windows allow in warmth and light (controlled by solar shades) as well as air.

Left: The elegant open-plan kitchen and living area proves that being green can go hand in hand with being stylish.

Below: Low-VOC paint has been used to great effect, as seen in the bathroom tucked behind the built-in kitchen units.

where the clients grow their own herbs and vegetables. It is such common-sense initiatives as these that have helped make this project a success. There is nothing particularly complicated or radical about this renovation; it is simply that the most sustainable solutions have been implemented at every juncture.

Although the clients received a 50 per cent state rebate towards the cost of the solar panels, Deam estimates that the sustainable materials added 15 per cent to the cost of the project. However, if there is a willingness to invest the extra time and expense, he believes that the materials and techniques used here could be applied to other homes. Sustainable renovation is an increasingly popular approach to green architecture in America, especially in the San Francisco Bay Area, as people are increasingly aware that it uses fewer building materials and less energy than demolition and rebuild. The Harris Residence is an inspirational example, and proves that sustainability and contemporary style can coexist.

TOTAL FLOOR AREA
430 square metres (4630 square feet)

ARCHITECT
Christopher C. Deam
99 Osgood Place
San Francisco
California 94133
USA
+1 415 981 1829
chris@cdeam.com
cdeam.com

PHOTOGRAPHY
Christopher C. Deam

HOUSE OF STEEL AND WOOD

RANÓN
SPAIN

Left: Supported by a steel frame, the house cantilevers dramatically out over the hillside.

Below: The architects clad the building in planks of locally sourced North pine and Douglas fir that vary in width, grain and colour, mimicking the appearance of trees' bark.

As the name suggests, this is a house constructed almost entirely of steel and wood. Its form draws on two aspects of traditional, vernacular Spanish architecture: the use of timber and the granary raised on stilts. But at the same time Madrid-based architects Ecosistema Urbano have thought carefully about how the building will affect its surroundings. The house now has little impact either visually or physically, and in the future it can be dismantled and recycled. It is a simple, economical design that uses very few materials, to great effect.

The client, a retired rugby player, had been living for a few years in an old *horreo* – a vernacular timber barn supported on stone legs – at the top of the site, in a forested valley near Ranón in the Asturias region of Spain. When the 36-square-metre (387-square-foot) building began to seem too small he decided to build a larger house in his sloping garden, locating it in an existing clearing among mature trees.

The new house provides nearly three times as much accommodation, with a floor area of 115 square metres (1240 square feet). The strength of its steel frame allows the entire building to be raised off the ground, maintaining the original gradient of the slope and allowing vegetation to grow back and cover the

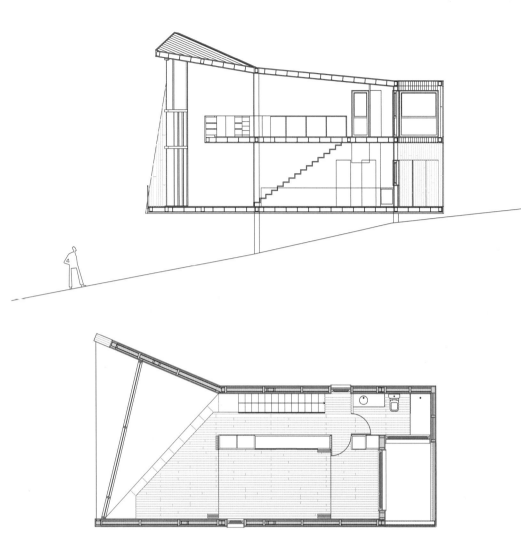

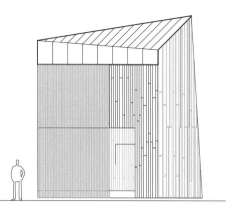

This page, clockwise from top left: Section; elevation; first-floor plan.

Opposite, top left: Wood has been used throughout the interiors, which have been designed to maximize space, with the ground-floor toilet concealed under the stairs.

Opposite, top right: As well as providing magnificent views across the valley, the south-facing glazed wall has opening windows set in timber frames.

Opposite, bottom: Behind the glazed wall, a double-height atrium is a key element in the passive cooling and heating of the house, allowing air to circulate freely between the ground and first floors.

scars of construction. The structure is essentially a simple rectangle, enabling standard steel and timber parts to be used throughout, minimizing costs and construction time. But as it cantilevers over the hillside, the eastern wall flicks out at an obtuse angle, expanding the surface area of the fully glazed south-facing wall and giving the building an added dynamism.

Most of the building, including the roof, is clad with planks of locally sourced North pine and Douglas fir, varying in width from 9 to 14 centimetres ($3\frac{1}{2}$ to $5\frac{1}{2}$ inches) and in length from 2.2 to 3.3 metres (7 to 11 feet). The various boards have been randomly dispersed across the surface of the building, the contrasting widths, grains, splits, colours and knots of the timber imparting an organic character. The overall effect resembles the bark of a tree and allows the building to merge into the wooded hillside.

The house is entered from the north through a porch protected from both the wind and prying eyes by a latticed timber screen. The interiors display a similar materials palette to the exterior, with North pine and Douglas fir appearing throughout. Even the staircase is beautifully crafted from timber boards. But it is the fully glazed southern wall that dominates the space, constantly pulling the eye towards the magnificent view. More than just a focal point, this glazed wall controls the temperature within the building. Operable windows set within timber frames have been inserted at points across its surface. In combination with the narrow windows on the east and west façades of the building, they open fully to allow ventilating breezes to sweep through the house in the summer. In the winter, the double-glazed panes allow radiant heat to enter the building, and the airtight timber frame wrapped with a 160-millimetre ($6\frac{1}{3}$-inch) layer of rock wool then traps this warmth. The double-height space directly behind the glazed wall facilitates the circulation of air throughout the house, encouraging warmer air to escape through high windows in the summer, and helping to distribute hot air through the building in the winter.

Inside the house is simple and minimalist. On the ground floor is an open-plan kitchen and living area, with a toilet concealed under the stairs. The first floor

was designed for maximum adaptability, with a family bathroom and a main space that can be converted into one, two or three bedrooms. This foresight on the part of the architects proved fortuitous, since during the project the client got married, and the couple are now adopting a child.

But this house also looks beyond its years of occupancy to the point when it will no longer be needed as a home. Its standard steel and timber parts and predominant use of wood for both interior and exterior finishes mean it can be easily dismantled and either rebuilt or recycled. It is a striking example of discreet and transient architecture that could one day disappear, leaving little evidence that it ever existed.

TOTAL FLOOR AREA
115 square metres (1240 square feet)

ARCHITECT
Ecosistema Urbano
Estanislao Figueras, 6
28008 Madrid
Spain
+34 (0)9 1559 1601
info@ecosistemaurbano.com
ecosistemaurbano.com

PHOTOGRAPHY
Emilio P. Doiztua

TOKYO SMALL HOUSE WITH THE SUN

TOKYO
JAPAN

Architect Satoshi Irei had three major objectives when designing this house in Tokyo's Nerima ward: to maximize space within the small site; to harness the sun's energy to heat the building; and to create a connection between the private areas of the house and the public street. He has achieved all three with a contemporary building that gives little clue to its sustainable agenda. Despite its modest appearance, this house contains some very progressive ideas.

The most revolutionary aspect of the building is located in the roof. The house has been fitted with the OM solar system – a passive solar-based system that uses air heated by the sun to control internal temperatures. Fresh air brought from the eaves into a narrow air passage in the roof is heated through roof panels made of tempered glass and metal, then channelled through an interior duct to a heat-storing

concrete slab beneath the ground floor. The slab directly warms the ground floor and releases hot air through floor vents distributed throughout the building's interior spaces. A wood-pellet boiler provides hot water and, on the coldest of days, auxiliary heating. In the summer, surplus hot air is redirected into the bathroom, where it is used to dry laundry. Owing to its thermally massive properties, the concrete slab also stores solar heat in the winter for release at night and retains coolness on summer nights for release during hot days. Irei is particularly pleased with the application of the OM system in this house, as he has been able to hide the duct within a plywood section of the wall, completely concealing the machinery from sight.

An equally important element of the design is the relationship it establishes between the private space

Opposite: At three storeys high, the house fully exploits its tight urban plot.

Below and right: The bamboo hedge at the front of the property only partially obscures views in from the street. The clever four-way sliding door allows the clients to vary the relationship between interior and exterior spaces.

of the occupants and the public domain of the street. In all his work the architect encourages metropolitan residents to connect with their surroundings, basing his designs on the open style of housing found in his native Okinawa prefecture, where sliding doors at the front of homes are often left open in the tropical heat. Such houses are surrounded by stone walls – *hinpun* – which block direct views in without entirely cutting off views out, encouraging interaction between occupants and passers-by. The Nerima house has no surrounding fences or walls; a black bamboo hedge acts as a *hinpun*, shielding a large door that opens into the house's main living area. The clients can precisely control their level of interaction with the world outside by choosing any of four sliding doors: a storm shutter, a mesh screen door, a glazed door and an internal *shoji* screen.

The house was designed to exploit its 30-square-metre (320-square-foot) footprint, and provides 60 square metres (640 square feet) of living area spread over three floors. Its timber frame is clad externally in galvanized sheets made from aluminium and steel with a corrugated detail that gives the house an unusual ribbed appearance. The walls and roof are insulated with 100 millimetres (4 inches) of glass wool. Inside there is an open-plan living, kitchen and dining area on the ground floor with an enclosed pantry and Japanese-style room containing *tatami* mats but no furniture. On the second floor is a bathroom, shower room and toilet, and a larger Japanese-style room where the owners sleep. From here a ladder leads to an open loft. Interior finishes have been kept minimal and neutral to create a feeling of light and space. Ceilings and walls of painted plasterboard contrast with a variety of woods: red pine for the floors (with *tatami* mats in the two Japanese-style rooms), southern yellow pine for the stairs, Japanese lime for built-in bookcases, and mahogany and spruce for the kitchen cabinets, as well as cedar and Douglas fir.

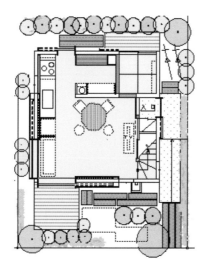

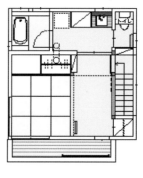

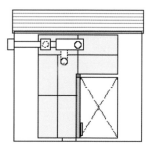

Irei believes that Tokyo is becoming an increasingly savage and crime-ridden place, and that this prompts residents to seal themselves away inside their homes. With this compact, elegant house he hopes to encourage people to interact more with their neighbours and to invest in forms of renewable energy. The aim is not only to sustain the planet's resources but also to rebuild the ties between public and private spaces that create strong, supportive communities.

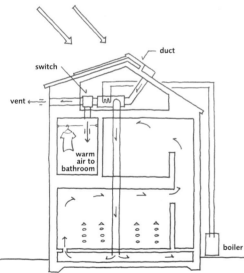

Opposite, top left: A ladder leads to the loft space above the family sleeping area.

Opposite, top right: The interiors are simply furnished and benefit from generous amounts of natural light. Here the four-way door has been closed using the shoji screen.

Opposite, bottom, from left: Site plan; first-floor plan; second-floor plan.

Above: Various species of timber have been employed inside the house to provide an understated elegance.

Left: The architect's diagram demonstrates how air heated by the OM passive solar system is distributed through the house.

SITE AREA
75 square metres (807 square feet)

TOTAL FLOOR AREA
60 square metres (640 square feet)

ARCHITECT
Satoshi Irei Architect & Associates
3-20-24 Mejiro
Toshimaku
Tokyo 170-0000
Japan
+81 (0)3 3565 7344
irei@interlink.or.jp; irei.exblog.jp

PHOTOGRAPHY
Satoshi Irei

TRIANGLE HOUSE

NESODDEN
OSLO
NORWAY

The Triangle House is a bold building with various chunks sliced from its corners adding to the dynamism of its sharp, angular walls and strict geometric lines. But this sculptural form is also a practical home for three generations of the same family. It was constructed from economical and sustainable materials and demonstrates how basic components can be manipulated to form an expressive and energy-efficient home.

The clients, a family of four, asked Oslo-based Jarmund/Vigsnæs Architects to create an economical and environmentally friendly home that capitalized on views of the nearby fjord and of Oslo, which lies 11 kilometres (7 miles) away. They needed a basement apartment for the children's grandmother and storage space for their enormous collection of books. The solution was a three-storey triangular form with walls precisely angled to capture vistas in all directions.

For the basement of the 286-square-metre (3080-square-foot) house, the architects chose concrete, which allowed them to cut into the slope behind the property and secure the building to the awkward plot, as well as introducing thermal mass into the structure. On top of this rests a lightweight timber frame. Timber construction is the most common building technique in the area, and both wood and labour are readily available locally. Wood is also cheap and quick to build with, and has the flexibility to create the cantilevers that are an integral part of this design. It also suits the low-energy aspirations of this building, since it is renewable and recyclable with

Page 184: The house's unusual shape was designed to capture views of nearby Oslo and the surrounding landscape.

Previous page: A concrete basement anchors the house to its sloping site, while the structure above is of lightweight timber – a traditional, local material that is cheap and renewable, and was easily adapted to the house's irregular shapes.

Left: The wall in the open-plan double-height living area functions as both a room partition and a vast set of built-in bookshelves.

The polished concrete floor in the kitchen is complemented by a concrete workbench designed to include a sink. Concrete floors throughout the building absorb solar gain, helping to heat the building.

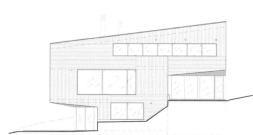

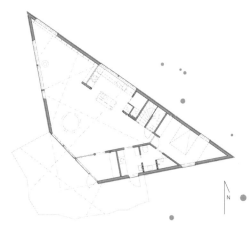

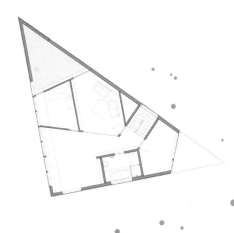

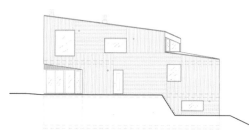

a low embodied energy. Once complete, the frame was clad in sustainable Siberian larch boards stained dark brown with iron vitriol and laid both horizontally and vertically to correspond to the window casements and add further interest to the façades.

In contrast to the focused approach of the exterior, where the large windows frame specific views, the living areas are fluid, with open and overlapping spaces. Above the self-contained basement flat is a bathroom, bedroom and laundry, but the majority of the ground floor is consumed by an open-plan kitchen, dining and living area, the latter a cavernous double-height space that fills the entire north-west point of the triangle. The top floor is more private, with two bedrooms, a guest room, a generous study and a family bathroom. Rooms

throughout the house are irregularly shaped, with mirrors in strategic positions to obscure any sense of conformity even further.

The architects exploited economical and sustainable materials to create warm and interesting interior surfaces. Orientated strand board (OSB) was sanded and sealed, providing a cheap and environmentally friendly equivalent to hardwood. It forms a skin over much of the interior, including walls, ceilings, stairwells, doors and even built-in bookshelves and furniture. These unbroken surfaces make the interior feel larger than it is and create a soft, natural ambience.

The floors throughout the building are cast in concrete, with lightweight concrete used in the upper levels. In the main living areas these floors are polished and left exposed, while in the bedrooms they are covered with sisal mats. This economical use of materials produces a hard-wearing and contemporary floor, but the concrete is also an essential element in the heating of this building. Radiant heat entering through the many windows is absorbed by this thermally massive material and gradually released throughout the day. Water pipes were laid under the concrete at every level to provide underfloor heating. This in turn is connected to a ground-source heat pump, which is submerged under the building and extracts latent heat from the ground to supply hot water. The house is also equipped with a wood-burning stove, although this holds greater importance as a symbolic hearth than as a heating device, since it is needed only during the coldest Norwegian days. A warm blanket of insulation traps the heat, with 200 millimetres (8 inches) of glass wool in the walls and 300 millimetres (12 inches) in the roof, and double-glazed windows fitted into airtight timber frames.

Throughout this project Jarmund/Vigsnæs Architects have shown an adventurous approach to both form and materials, demonstrating that an energy-efficient home can be both sculptural and practical. By stripping the building of all unnecessary finishes, the architects have shown how economical construction materials, which are usually kept hidden, can be transformed into arresting and contemporary surfaces. Cheap to build and cheap to run, the Triangle House is a forceful building, both in terms of its appearance and in the lesson it provides for sustainable architecture.

TOTAL FLOOR AREA
286 square metres (3080 square feet)

ARCHITECT
Jarmund/Vigsnæs Architects
Hausmannsgate 6
0186 Oslo
Norway
+47 (0)22 99 43 43
jva@jva.no
jva.no

PHOTOGRAPHY
Ivan Brodey

Opposite, left (from top): West elevation; ground-floor plan; first-floor plan; south elevation.

Opposite, right: Orientated strand board (OSB) – a cheap and environmentally friendly material made from leftover woodchips – was used extensively for walls, ceilings and even stairwells.

Above: Hot water is provided by a ground-source heat pump, which absorbs latent heat from the ground.

Right: The exterior cladding of stained Siberian larch is laid both vertically and horizontally to add further dynamism to the crisp triangular design.

SWISS AMBASSADOR'S RESIDENCE

WASHINGTON, D.C.
USA

The residence of the Swiss Ambassador in Washington is both a private home and a public space for official functions. Situated in the smart residential district in the north-west of the city, the building commands an impressive view of the Washington Monument and has been designed in a cruciform shape to capitalize on this outlook. It is an imposing building with contrasting charcoal-grey and translucent white walls that remind guests of the snow-capped mountains of Switzerland. It is also a building that reflects the country's progressive attitude towards sustainable construction. Architects Steven Holl and Justin Rüssli have together created a light, spacious and low-energy home through which the Swiss ambassador can demonstrate to his guests how elegant ecological solutions can be.

The residence is positioned on a plateau with four courtyards of varying character placed in each angle of the cross's arms: an arrival square, a pool that reflects dappled light on to the building's exterior, an outdoor reception space with views across the city, and a more private herb garden. Each square is connected to the ground floor of the building, which is dedicated to public areas, including two formal

With its bold lines, sharp edges and grey-and-white surfaces, the building draws inspiration from the snow-capped mountains of Switzerland and affirms Swiss environmental awareness, with, for example, a green roof that helps cut air pollution in this urban environment.

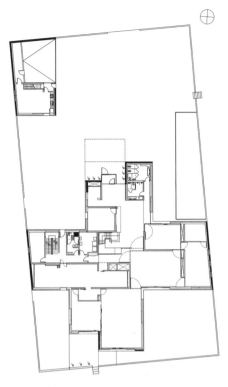 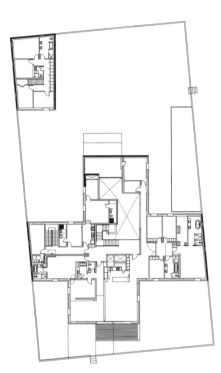

dining-rooms, three interlocking salons and a grand double-height entrance hall. Large sliding doors allow the ambassador to change the configuration of these rooms. Upstairs are the private areas: a generous apartment for the ambassador as well as staff rooms and two guest rooms. It is a gigantic 2140 square metres (23,000 square feet) in total.

The house has been constructed to meet the Swiss Minergie standard and is therefore exceptionally well insulated and energy efficient. The key materials are concrete, stained charcoal-grey and textured by the shutters in which it was set, and structural glass planks, laid over transparent double glazing in some sections and backed with opaque thermal insulation in others. The visual impact of these materials is striking – at night the house glows from within – and their thermal properties are augmented by an 800-millimetre-thick (31½ inches) layer of insulation in the walls throughout.

The building's green roof adds to this thermal blanket and prevents water run-off. It also reduces the 'heat-island effect', whereby large urban areas can

Opposite, top: Ground- (left) and first-floor plans. A caretaker's house is separate from the main building. The main entrance area is at the bottom left of the ground-floor plan; it leads into the entertaining spaces and the terrace. Guest and staff quarters are in the left-hand wing on the first floor.

Opposite, bottom: The cruciform shape of the building boosts energy efficiency by maximizing natural light and passive solar gain, and allows for a courtyard of varying function in each corner of the cross's arms.

Above and right: The external envelope of charcoal-grey-stained textured concrete and contrasting structural glass planks not only provides the benefits of thermal mass but also brings visual and tactile qualities that allow the building to appear imposing without being overtly bulky.

be as much as 5°C (9°F) hotter than the surrounding countryside, partly because hard, reflective surfaces such as roofs and roads absorb solar energy and radiate this heat back into the atmosphere when air temperatures fall. Plants on green roofs absorb latent heat energy and use it for evapotranspiration – the process by which water moves through a plant and is eventually released from the leaves as water vapour – reducing urban temperatures and cutting air pollution and smog. In large residential projects such as the Swiss Ambassador's Residence, therefore, a green roof makes a considerable contribution to improving the surrounding air quality.

Installed among the plants on the roof are photovoltaic panels to help meet the building's electricity demands, which are already reduced through the use of as much natural light as possible. The southern façades are designed to maximize passive solar gain, and digitally operated external sunshades control the amount of radiant heat entering the house. The open, lofty spaces inside the building can also be naturally cross-ventilated via strategically placed doors.

Plastered, white-painted walls keep the interiors minimalist. Additional finishes have been selected for sustainability: visitors to the public areas walk on terrazzo speckled with pieces of recycled glass, and upstairs the floors are laid with sustainable bamboo. It is an ethereal space that successfully combines a flexible public area with a comfortable home.

With its commanding location and sophisticated design, the Swiss Ambassador's Residence is an impressive architectural expression of Switzerland's cultural and environmental aspirations. In its capacity as an embassy this building will welcome more guests in a year than most houses do in a lifetime, making the structure itself an ambassador for environmentally responsible design.

Right: The interiors are simple and minimalist, with floors of terrazzo and recycled glass in the public areas downstairs and sustainable bamboo flooring in the private residential areas upstairs.

Opposite, left: On the first floor, private rooms are arranged around the double-height entrance hall.

Opposite, right: Large sliding doors allow the public spaces to be carved up into numerous configurations.

TOTAL FLOOR AREA
2140 square metres (23,000 square feet)

ARCHITECTS
Steven Holl Architects
450 West 31st Street, 11th Floor
New York, New York 10001
+1 212 629 7262
nyc@stevenholl.com
stevenholl.com

Rüssli Architekten
Neustadtstrasse 3
CH-6003 Lucerne
Switzerland
+41 (0)41 226 21 81
mail@ruessli.ch; ruessli.ch

PHOTOGRAPHY
Pages 190–91: Prakash Patel
Pages 192–95: Andy Ryan

THE HOME

CLOONE
COUNTY LEITRIM
REPUBLIC OF IRELAND

The Home in Cloone is constantly evolving. Designed by architect Dominic Stevens and his wife, Mari-Aymone Djeribi, it is an unpretentious building that can adapt to their young family's needs. They built the house themselves using local materials and low-tech and affordable building methods, focusing every decision on creating a sustainable home that can in turn support a sustainable lifestyle. Their all-encompassing approach highlights the advantages of vernacular materials and techniques and has fostered strong bonds with the surrounding land and the local community.

To exploit the natural benefits of the site, the couple began by living there in tents with a makeshift outdoor kitchen, giving them the opportunity to observe sun and wind patterns, views and routes across the site. Forty-four days later they had constructed two weatherproof timber boxes – the backbones of the house – into which they moved first the kitchen and then slowly themselves. Living in the boxes, they gained a true sense of the space, which they gradually carved up according to function by erecting partition walls. A straw-bale extension was then added to the north side of the boxes, linking them. Stevens and Djeribi say that as their lives continue to alter, so too will their home.

The different building techniques used in the 200-square-metre (2150-square-foot) building define two different kinds of space. The two horizontal timber boxes, built at right angles to one another, are divided into simple rooms that house specific functions, such as cooking, eating and sleeping. Prefabricated on site in the couple's polytunnel workshop, the frames for these boxes

were inspired by the cheap, easily constructed, timber-framed housing pioneered by Walter Segal in Britain in the 1960s and 1970s. The dimensions of the boxes are based on standard 1.2 × 2.5-metre (4 × 8-foot) timber sheet sizes, allowing the couple to make last-minute decisions when it came to construction without wasting prefabricated parts, and facilitating easy reconfiguration in the future. The southern façades of the boxes are almost entirely glazed. The exterior cladding is Sitka spruce, which is usually used as pallet wood; at the end of its lifespan of seven to ten years, the spruce will be burnt in the house's stove and the building fitted with a fresh skin.

The L-shaped straw-bale structure that connects the two timber boxes has no official function, providing an ambiguous place to loiter undisturbed in the shelter of its thick, earthy walls. The roof was erected on concrete load-bearing walls before the straw bales arrived, eliminating the need for a temporary shelter to keep the bales dry during construction. The bales act as non-structural screen walls strapped to the concrete blocks at intervals. They have been rendered inside and out with a thick lime plaster that gives the building a warm, homely feel. Inside, the floors follow the gradient of the ground, combining with the thick, irregular walls to create the ambience of an excavated cave.

As well as providing contrasting spatial experiences, the two different structures create a very energy-efficient building. In the summer large openings on the west and east elevations admit gentle breezes to cool the house. In the winter the extensive south-facing glazing admits sunlight that heats the thermally massive north side of the building. Thick insulation in

Opposite: The Home has been gradually self-built in response to its owners' needs, resulting in a building of simple form but rich materiality.

Below: Looking from the west, the two contrasting techniques used to build the house are clearly visible (left). The southern façade of the timber boxes is glazed (top right). The contrasting L-shaped cob structure to the north has thick earthy walls and minimal openings (bottom right).

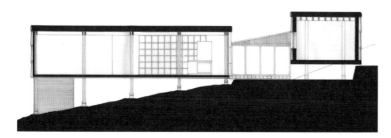

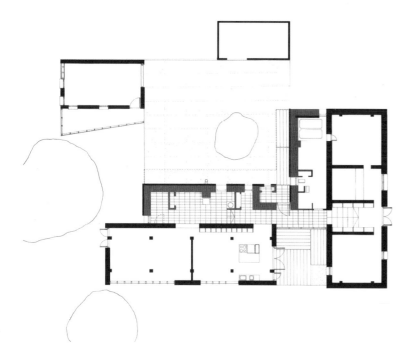

the walls of the timber structure and in the roof guards against excessive heat loss, and additional warmth is supplied by a wood-burning stove.

The couple have planted trees along existing hedgerows; they will coppice them (cut them down to ground level periodically and harvest the regenerating branches) and become self-sufficient in fuel within ten years. They are also farming the 1.6 hectares (4 acres) of land around the house, raising chickens, ducks and goats for eggs, meat and milk. They grow vegetables and fruit, swapping anything they do not consume for alternative goods with their neighbours.

Until recently self-sufficiency and a strong community network would have been the norm in Ireland. Building a home, for example, was a vernacular tradition that called on many local people to lend a helping hand. Through The Home in Cloone and the lifestyle they lead there, Stevens and Djeribi are following these traditions, relying not only on local and renewable resources for their food, fuel and shelter but also on the interdependence that brings communities together. The Home looks completely different from the hundreds of developer-built houses currently spoiling the Irish countryside. Designed as a home for life, not as a commodity to be bought and sold, it is a reminder that lessons from the past could hold the key to the future.

Opposite, left: Section (top) and site plan.

Opposite, right: Only after they had lived for a while in the timber boxes did the owners partition the space to define various uses.

Left: The interiors are simply decorated and sparsely furnished; the structural bones of the timber boxes have been left exposed.

SITE AREA
1.6 hectares (4 acres)

TOTAL FLOOR AREA
200 square metres (2150 square feet)

ARCHITECT
Dominic Stevens Architects
Annaghmaconway
Cloone
County Leitrim
Republic of Ireland
+353 (0)71 963 6988

PHOTOGRAPHY
Ros Kavanagh

Glossary

Alucobond
Two sheets of aluminium thermobonded to a polyethylene core. Has a high embodied energy but is made up of 85 per cent recycled material.

Ambient cooling/heating
A change in air temperature produced by non-mechanical means.

Biomass boiler
A boiler that burns organic material rather than fossil fuels. The biomass comes either directly from plants (for example, wood) or indirectly from industrial, commercial, domestic or agricultural products (for example, animal waste, biodegradable products from the food-processing industry and such high-energy crops as rape and sugar cane). For small-scale domestic appliances biomass usually takes the form of wood pellets, wood chips or logs.

Caisson
A retaining watertight structure that allows water to be pumped out during construction, keeping the working environment dry.

Cantilever
A load-bearing beam supported at one end only, allowing the construction of overhanging structures without external bracing.

Carbon store/carbon sink
A concept based on the natural ability of plants and soil to soak up carbon dioxide from the atmosphere and store that carbon during their lifetime. Trees, for example, absorb carbon (as carbon dioxide) and release oxygen during photosynthesis. That carbon is released back into the atmosphere as carbon dioxide only once the timber is burned. The concept of carbon sinks or stores has become more widely known because the Kyoto Protocol allows the use of carbon dioxide sinks as a form of carbon offset.

Cellulose insulation
A type of insulation material made from plant fibre, commonly recycled paper.

Chipboard
An engineered wood product made from woodchips, sawmill shavings and sawdust or recycled wood particles, mixed with a synthetic resin, pressed and extruded.

Composting toilet
Any sanitary system that converts human waste into compost or usable soil through the natural breakdown of organic matter into its essential minerals.

Concrete, pre-cast or cast *in situ*
A construction material that consists of cement (commonly Portland cement) and other materials (such as fly ash and slag cement), aggregate (generally a coarse aggregate such as limestone or granite, plus a fine aggregate such as sand), water and chemical admixtures. The most ecological concrete is that created with a percentage of aggregate made from pulverized fuel ash or ground, granulated blast-furnace slag, both by-products of power stations. Crushed pre-existing concrete can also be used as an aggregate. Pre-cast concrete is shaped in a factory; concrete cast *in situ* is generally poured between wooden shutters, which are removed after it has set, or between permanent insulating boards.

Cross-ventilation
The cooling of internal spaces using natural breezes.

CSA
Canadian Standards Association

Double-flux ventilation system
A heat-exchange system that constantly pumps stale air out of a building and fresh air in. In the winter heat from the air being expelled is transferred to the air entering the building. In the summer the reverse process occurs, with the incoming air cooled by that leaving the building.

Drywall
Also known as plasterboard, wallboard or gypsum board, drywall is made from plaster, which is then covered on both sides by fibreglass matting or heavyweight paper. Sheets of drywall are attached to studwork and ceiling joists to finish the interior walls and ceilings of a building. Because it is a dry process this method is much faster and easier than the traditional application of wet plaster.

Dual-flush toilet
A toilet that has both a light and a heavy flush option, and uses less water than a conventional model. Conventional toilets can flush as much as 13 litres (3 gallons) of water, but a dual-flush toilet will use only 2–3 litres (½–⅔ gallon) on a short flush and 4–6 litres (1–1½ gallons) on a full flush. Now obligatory in some countries.

Ecoply
Structural plywood panels made from Radiata pine grown in sustainable plantations.

Embodied energy
The amount of energy consumed in the acquisition of raw materials, their processing and the manufacturing, transportation to site and construction to the point of use of a product or material.

Envelope
The outer skin of a building.

Flax insulation
A natural, renewable and non-toxic insulation material made from fibres extracted from the outer layer of the stem of a flax plant.

Foamglas insulation
An insulation material consisting of minute sealed glass cells formed by chemically reacting finely ground oxidized glass with carbon at a high temperature. It has a high embodied energy, but over 60 per cent of the glass used comes from post-consumer waste.

FSC
Forest Stewardship Council.

GGBS (ground, granulated blast-furnace slag)
A by-product of iron production commonly used to supplement Portland cement in the production of concrete.

Glulam (glue-laminated) timber
A structural product manufactured by gluing together individual pieces of timber. The resulting beams are stronger than similarly sized pieces of solid wood and can span greater distances.

Ground-source heat pump
A system to extract latent heat from the ground via a borehole or network of underground pipes, for domestic heating and hot water.

Heat bridge
Areas of materials with high thermal conductivity, which allow heat to leak easily from a building.

Heat-recovery system
An air-handling system that collects and exhausts stale air from the inside of a building while drawing in and distributing fresh air at the same temperature. Reduces the need for mechanical heating and cooling.

Heat sink
A material, such as concrete, that absorbs and dissipates heat, helping to regulate the internal temperature of a building.

Homasote
An acoustic panel made from recycled newspaper.

Lime plaster
A non-toxic plaster made from processed lime, water and aggregate.

Low-emissivity (low-e) coating
A microscopically thin metallic layer deposited directly on to the surface of one or more of the panes of glass in a window. Improves thermal performance by reducing heat flow between the panes.

Misapor concrete
A Swiss product made with a recycled Foamglas aggregate. The Foamglas produces a dense honeycomb structure, which traps air within the concrete. These air pockets do not affect the strength or durability of the concrete but reduce its weight and act as built-in insulation.

Orientated strand board (OSB)
An engineered wood product formed by compressing wood chips with high-strength adhesive. When sanded, stained and sealed the OSB panels provide a cost-effective and responsible alternative to hardwood boards.

PassivHaus principles
A set of standards developed in Germany to create ultra-low-energy buildings. A similar standard, Minergie-P, is used in Switzerland.

Patina
A coating of various chemical compounds, such as oxides or carbonates, that forms on a metal surface when it is exposed to the weather.

Pavatherm insulating fibreboards
Rigid boards with thermal and acoustic insulating properties, made from wood pulp. No adhesives or wood preservatives are used in their manufacture.

PEFC
Programme for the Endorsement of Forest Certification.

PFA (pulverized fly ash)
A residue from the combustion of coal. Commonly used to supplement Portland cement in the production of concrete.

Photovoltaic (PV) panel
A panel consisting of layers of semi-conducting materials that generate electrical charges from sunlight to supply electrical current.

Plywood

A type of engineered wood made from thin sheets of timber veneer glued together. Each sheet is glued with its grain at right angles to the adjacent layer to increase the strength of the product.

Polyester bulk insulation

Insulation made from polyester (a man-made fibre comprising chains of synthetic carbon polymers, also used in pillows, mattresses and clothing).

Porous paint

Paint through which air and moisture can pass, allowing a wall to breathe.

Radiant heat

Radiant energy emitted from heat sources including the sun. It can be absorbed by other objects, for example solar thermal hot-water panels and construction materials such as concrete.

Rammed-earth construction

The process of building with compressed earth. The soil must contain suitable proportions of clay, sand and aggregate, and is set in forms, which are removed once the walls have hardened.

Renewable energy

Energy created from the earth's natural and replenishing resources, such as the sun, wind, tides and geothermal heat (underground heat).

Rock wool

An insulation material made primarily from basalt, a volcanic rock.

SFI

Sustainable Forestry Initiative

Solar gain

Obtaining energy from the sun. Most often used to describe the heating of a space by sunlight through glass.

Solar thermal hot-water panel

A means of harvesting the sun's energy to provide hot water. The various panel systems contain fluid, which is heated by the sun; the heat is then transferred to the domestic water supply.

Stack effect

The movement of air into and out of a building, driven by buoyancy created by a difference in indoor-to-outdoor air temperatures. Cooler air is heavier and warmer air lighter, so the greater the thermal difference and the height of the building, the greater the buoyancy force and the stronger the stack effect.

Thermal mass

The capacity of a material to store and release heat. Concrete, for example, has a high thermal mass, as it slowly absorbs or releases relatively large quantities of heat per unit volume compared to other materials, such as glass. When used correctly in construction it can significantly reduce the need for mechanical heating and cooling.

Thermobonding

The use of high temperatures to bond composite fibres together permanently.

Thermosetting resin

A material that hardens when heated and cannot be remoulded.

ThermoWood

An exceptionally durable timber made by exposing softwood to temperatures in excess of 200°C (390°F) to drive out moisture and resin. Sourced from sustainable forests and approved by the PEFC.

Trespa

An extremely durable material made from thermosetting resins reinforced with wood fibres, manufactured at high pressure and temperature. Designed to breathe, therefore preventing damp and rot, and made using timber from sustainable forests. Produces no harmful gases when burned.

UltraTouch cotton insulation

An insulation material made from mulched denim.

VOCs (volatile organic compounds)

Airborne particles that give off a strong smell and contribute to poor air quality, both indoors and out. Found in many household products, including paints, varnishes and adhesives. VOCs include the chemicals xylene, benzene and toluene and are toxic to the nervous system; some have been linked with cancer.

Wind scoop

A funnel used to force wind into a specific space. Most commonly used on boats to force air through a hatch to ventilate the areas below deck, wind scoops can also be attached to buildings to help naturally ventilate interior spaces.

Further Reading

Books

Will Anderson, *Diary of An Eco-Builder*, Dartington, UK (Green Books) 2006

Alistair Fuad-Luke, *The Eco-Design Handbook: A Complete Sourcebook for the Home and Office*, London (Thames & Hudson) 2004

James Grayson Trulove, *Sustainable Homes: 26 Designs That Respect the Earth*, New York (HarperCollins) 2004

Leo Hickman, *A Good Life, the Guide to Ethical Living*, London (Eden Project Books) 2008

Alanna Stang and Christopher Hawthorne, *The Green House: New Directions in Sustainable Architecture*, New York (Princeton Architectural Press) 2005

Magazines and journals

A10: a10.eu
AIT: ait-online.de
Architects' Journal: ajplus.co.uk
Architectural Record: archrecord.construction.com
Architectural Review: arplus.com
Domus: domusweb.it
Dwell: dwell.com
The Ecologist: theecologist.org
Environmental Building News: buildinggreen.com
Grand Designs: granddesignsmagazine.com
Green Futures: forumforthefutures.org.uk
Japan Architect: japan-architect.co.jp
New Consumer: newconsumer.com

Websites

Canadian Standards Association: csa.ca
Forest Stewardship Council: fsc.org
Programme for the Endorsement of Forest Certification: pefc.org
Sustainable Forestry Initiative: sfiprogram.org
Sustainable Life: sustlife.com
Treehugger: treehugger.com
Wood for Good: woodforgood.com
World Architecture News: worldarchitecturenews.com

UK and Europe

AEE (Institute for Sustainable Technologies; Austria): aee-intec.at
Association HQE (France): assohqe.org
Centre for Alternative Technology (UK): cat.org.uk
Construction Resources (UK): constructionresources.com
Energy Research Centre of The Netherlands: ecn.nl
Energy Saving Trust (UK): energysavingtrust.org.uk
Greenphase (environmental directory): greenphase.com
GreenSpec (UK): greenspec.co.uk
LivingRoofs: livingroofs.org
Minergie (Switzerland): minergie.com
Passiefhuis Platform (Belgium): passiefhuisplatform.be
PassivHaus Institute (Germany): passiv.de
PassivHaus UK: passivhaus.org.uk
Promotion of European Passive Houses: europeanpassivehouses.org
Timber Research and Development Association (UK): trada.co.uk
UK Green Building Council: ukgbc.org

North America

Alliance to Save Energy: ase.org
American Solar Energy Society: ases.org
Architects, Designers, Planners for Social Responsibility: adpsr.org
Canadian Office of Energy Efficiency: oee.nrcan.gc.ca
FabPrefab: fabprefab.com
Healthy Building Network: healthybuilding.net
Livable Places: livableplaces.org
Northwest Ecobuilding Guild: ecobuilding.org
Sustainable Architecture, Building and Culture: sustainableabc.com
Sustainable Communities Network: sustainable.org
Sustainable Products Corporation: sustainableproducts.com
US Department of Energy, Energy Efficiency and Renewable Energy: eere.energy.gov
US Green Building Council: usgbc.org

Australia and New Zealand

Australian Conservation Foundation: acfonline.org.au
Australian Government, Department of the Environment, Water, Heritage and the Arts: environment.gov.au
Building Biology and Ecology Institute (New Zealand): ecoprojects.co.nz
Eco Directory: ecodirectory.com.au
Eco Property and Resources: eco.com.au
Guide to choosing energy-efficient appliances: energyrating.gov.au
Society for Responsible Design: srd.org.au
Sustainable Living Foundation: slf.org.au

Picture Credits

© 8Inch (eightinch.co.uk): 41l; © Peter Aaron/Esto/View: back jacket br, 19r, 44–49; © ADM Systems (admsystems.co.uk): 25t; © All About Bricks (allaboutbricks.co.uk): 16t; © The Alternative Flooring Company (alternativeflooring.com): 26b; © American Cork Products (amcork.com): 28r; © Photography courtesy of American Standard, AS America, Inc. (americanstandard-us.com): 36l; © Peter Bennetts: 68–73; © Patrick Blanc: 135; © Louise Body (louisebodywallprint.com): back jacket tc, 31b; © Ivan Boiko: 150br, 151l; © Ivan Brodey: 184–89; © Bye Bye Standby (byebyestandby.co.uk): 34b; © Caroma (caroma.com.au): 37br; © Peter Cook/View: 158–61; © Christopher C. Deam: front jacket, 170–75; © Emilio P. Doiztua: 176–79; © drummonds-arch.co.uk: 37t; © EC Power (ecpower.co.uk): 25b; © Richard Fernau/Fernau Hartman: 94, 95t, 97t, 98, 101bl; © Forbo (forbo-flooring.com): 28l; © GEC Anderson (gecanderson.co.uk): 40c; © Florian Golay: 132–33, 134t; © John Gollings: 142, 144–47; © The Green Shop (greenshop.co.uk): 30t, b; © Greenhaus (greenhaus.co.uk): 32tr; © Art Grice: 112–17; © Andrew Griffiths/lensaloft.com: 143; © Grohe (grohe.co.uk): 36r; © Roland Halbe: 6–7, 136–41; © Stuart Haygarth (stuarthaygarth.com): 33l; © Shu He: 167–69; © Hiroyuki Hirai: 152–57; © Tim Hursley: 99bl; © Peter Hyatt: 84–87; © Ibstock Brick Ltd (ibstock.com): 15br; © Ice Energy (iceenergy.co.uk): 22b; © Oliver Ike/Ike Branco: back jacket bl, 56–57, 58b, 59b, 60–61; © Satoshi Irei: 180–83; © Thomas Jantscher: 8–9, 17r, 27l, 126–31; © Photography by Rob Judges (rob@robjudges.com), architect Roderick James Architects LLP, oak frame by Carpenter Oak Ltd: back jacket tcr, 14l, 39r; © David Juet: 118–21; © Ros Kavanagh: 196–99; © G.G. Kirchner: 102–105; © Gunnar Knechtel: 20t, 164, 165tr, b; © Paul Kozlowski: 23, 62–67;

© Kim Yong Kwan: 74–77; © Kevin Lake: 50; © J.K. Lawrence: 95b, 96b, 99tl, r, 100, 101tl, tr; © bamboo kitchen from Neil Lerner (neillerner.com): 38r; © Lime Technology (limetechnology.co.uk): 29t; © Kevin Low: 59t; © M-House (m-house.org): 18; © Milestone (milestone.uk.net): 41c; © Porcelanosa (porcelanosa.co.uk): 28c; © Ben Rahn/A-Frame/Taylor Smyth Architects: 122–25; © rainwaterharvesting.co.uk: 21b; © Marvin Rand, courtesy of Pugh + Scarpa Architects: 10–11, 40r, 78–83; © Mascoat (mascoat.com): 31t; © Nigel's Eco Store (nigelsecostore.com): 34t, 39l; © Prakash Patel: 190–91; © Project Russia: 148–149, 150bl, 151r; © Aniol Rescloso: 163; © The Rubber Flooring Company (therubberflooringcompany.co.uk): 27c; © Enric Ruiz-Geli: 162, 165tl; © Andy Ryan: 192–95; © Scrapile (scrapile.com): 33r; © Second Nature Kitchen Collection (sncollection.co.uk): 40l; © James Silverman: 88–89, 92–93; © Abey Smallcombe (abeysmallcombe.com): back jacket tl, 15bl; © Jefferson Smith/FLACQ Architects/Media10 Images: back jacket bc, 51–55; © Tim Soar/Velfac, architect Richard Pain: 19l; © Solar Century (solarcentury.co.uk): 22t; © Solid Floor (solidfloor.co.uk): back jacket tr, 26t; © steko.com: 14r; © Stonell (stonell.com): 27r; © Suffolk Housing Society: 17l; © Edmund Sumner/VIEW: 15t; © Thermafleece (secondnatureuk.com): 21t; © Jussi Tiainen: 106–11; © Tierrafino (tierrafino.com): 29b; © Ting (tinglondon.com): 32br; © The Emily Todhunter Collection by O Ecotextiles Inc: 32l; © Toto (totousa.com): 37bl; © Trunk Reclaimed (trunkreclaimed.co.uk): 38l, 41r; © Umicore (vmzinc.co.uk): 20b; © Vaillant (vaillant.co.uk): 35t; © Warmup Proformat (warmup.com): 35b; © Westfire: 24t; © Jim Wilson/New York Times/Eyevine: 97b, 101br; © Windsave Ltd: back jacket tcl, 24b; © Gert Wingårdh: 90–91; © Catherine Wonek/strawbalecentral.com: 16b; © Fu Xing: 166a

Index

Published by

Merrell Publishers Limited
81 Southwark Street
London SE1 0HX

merrellpublishers.com

First published 2008
Paperback edition first published 2010

British Library Cataloguing-in-Publication data:
Strongman, Cathy.
   The sustainable home : the essential guide to eco building, renovation and decoration.
   1. Architecture, Domestic – Environmental aspects.
   2. Sustainable architecture.
   I. Title
   728.3'7'047-dc22

ISBN 978-1-8589-4518-7

Produced by Merrell Publishers Limited
Designed by Martin Lovelock
Picture-researched by Helen Stallion
Copy-edited by Philippa Baker
Proof-read by Isabella McIntyre
Indexed by Hilary Bird
Printed and bound in China

Front cover: Harris Residence, Belvedere, California, USA (see pages 170–75)
Back cover, bottom left to right: Safari Roof House, Kuala Lumpur, Malaysia
   (see pages 56–61); Rowe Lane House, London, UK (see pages 50–55); Loblolly House,
   Taylors Island, Maryland, USA (see pages 44–49)
Pages 6–7: Casa en La Florida, Madrid, Spain (see pages 136–41)
Pages 8–9: Renovated House, Chamoson, Switzerland (see pages 126–31)
Pages 10–11: Solar Umbrella, Venice, California, USA (see pages 78–83)

Acknowledgements

I should like to thank the Merrell team, whose
support and endless enthusiasm have made
The Sustainable Home a joy to work on. In
particular, I am grateful to Julian Honer for
commissioning the book; Martin Lovelock and
Nicola Bailey, who were both instrumental in its
design; Nick Wheldon for picture coordination; and
freelance picture researcher Helen Stallion, who
miraculously sourced hundreds of images just in
time for the birth of her baby. A special thank you
goes to Rosanna Fairhead, who not only skilfully
project-managed the book, but also offered
much-appreciated encouragement throughout.

   I am also grateful to all the architects who
provided the information, drawings and photographs
that made this book possible. Their innovative
approaches to sustainable architecture deserve
the highest of praise. Thank you also to Antonio
Limones and Helen Strongman, both of whom
translated foreign texts. Finally, I should like to
thank my parents, and Duncan Donaldson for
putting up with the neuroses of a freelance writer
and for encouraging me to aim high.

Cathy Strongman, formerly Eco Editor at
Insideout magazine, published by the *Sunday
Times* newspaper, is now a freelance journalist. She
contributes articles to various publications, including
the *Architects' Journal*, *Grand Designs Magazine*
and *House & Garden*. Her first book, *New London
Architecture 2*, co-authored with architecture critic
Kenneth Powell, was published by Merrell in 2007.